INTRODUCTION TO THE
CALCULUS OF
VARIATIONS

2nd Edition

INTRODUCTION TO THE
CALCULUS OF
VARIATIONS

2nd Edition

BERNARD DACOROGNA
Ecole Polytechnique Fédérale, Switzerland

Imperial College Press

Published by

Imperial College Press
57 Shelton Street
Covent Garden
London WC2H 9HE

Distributed by

World Scientific Publishing Co. Pte. Ltd.
5 Toh Tuck Link, Singapore 596224
USA office: 27 Warren Street, Suite 401-402, Hackensack, NJ 07601
UK office: 57 Shelton Street, Covent Garden, London WC2H 9HE

Library of Congress Cataloging-in-Publication Data
Dacorogna, Bernard, 1953–
    [Introduction au calcul des variations. English]
    Introduction to the calculus of variations / by Bernard Dacorogna. -- 2nd ed.
        p. cm.
    Includes bibliographical references and index.
    ISBN-13: 978-1-84816-333-1 (hardcover : alk. paper)
    ISBN-10: 1-84816-333-9 (hardcover : alk. paper)
    ISBN-13: 978-1-84816-334-8 (pbk. : alk. paper)
    ISBN-10: 1-84816-334-7 (pbk. : alk. paper)
    1. Calculus of variations.   I. Title.
    QA315.D3413 2009
    515'.64--dc22

                                                        2008038721

British Library Cataloguing-in-Publication Data
A catalogue record for this book is available from the British Library.

Printed in Singapore.

I would like to thank all students and colleagues for their comments on the French version, in particular O. Besson and M. M. Marques who commented in writing. Ms. M. F. DeCarmine helped me by efficiently typing the manuscript. Finally my thanks go to C. Hebeisen for the drawing of the figures.

the next ones. It is much used in Chapters 3 and 4 but less in the others. All of them, besides numerous examples, contain exercises that are fully solved in Chapter 7.

Finally I would like to thank the students and assistants that followed my course; their interest has been a strong motivation for writing these notes. I would like to thank J. Sesiano for several discussions concerning the history of the calculus of variations, F. Weissbaum for the figures contained in the book and S. D. Chatterji who accepted my manuscript in his collection at Presses Polytechniques et Universitaires Romandes (PPUR). My thanks also go to the staff of PPUR for their excellent job.

Euler and Lagrange who found a systematic way of dealing with problems in this field by introducing what is now known as the Euler-Lagrange equation. This work was then extended in many ways by Bliss, Bolza, Carathéodory, Clebsch, Hahn, Hamilton, Hilbert, Kneser, Jacobi, Legendre, Mayer, Weierstrass, just to quote a few. For an interesting historical book on the one dimensional problems of the calculus of variations, see Goldstine [55].

In the nineteenth century and in parallel to some of the work that were mentioned above, probably, the most celebrated problem of the calculus of variations emerged, namely the study of the Dirichlet integral; a problem of multiple integrals. The importance of this problem was motivated by its relationship with the Laplace equation. Many important contributions were made by Dirichlet, Gauss, Thompson and Riemann among others. It was Hilbert who, at the turn of the twentieth century, solved the problem and was immediately after imitated by Lebesgue and then Tonelli. Their methods for solving the problem were, essentially, what are now known as the direct methods of the calculus of variations. We should also emphasize that the problem has been very important in the development of analysis in general and more notably functional analysis, measure theory, distribution theory, Sobolev spaces or partial differential equations. This influence is studied in the book by Monna [77].

The problem of minimal surfaces has also had, almost at the same time as the previous one, a strong influence on the calculus of variations. The problem was formulated by Lagrange in 1762. Many attempts to solve the problem were made by Ampère, Beltrami, Bernstein, Bonnet, Catalan, Darboux, Enneper, Haar, Korn, Legendre, Lie, Meusnier, Monge, Müntz, Riemann, H.A. Schwarz, Serret, Weierstrass, Weingarten and others. Douglas and Rado in 1930 gave, simultaneously and independently, the first complete proof. One of the first two Fields medals was awarded to Douglas in 1936 for having solved the problem. Immediately after the results of Douglas and Rado, many generalizations and improvements were made by Courant, Leray, MacShane, Morrey, Morse, Tonelli and many others since then. We refer for historical notes to Dierkes-Hildebrandt-Küster-Wohlrab [39] and Nitsche [82].

In 1900 at the International Congress of Mathematicians in Paris, Hilbert formulated 23 problems that he considered to be important for the development of mathematics in the twentieth century. Three of them (the 19th, 20th and 23rd) were devoted to the calculus of variations. These "predictions" of Hilbert have been amply justified all along the twentieth century and the field is at the turn of the twenty first one as active as in the previous century.

Finally we should mention that we will not speak of many important topics of the calculus of variations such as Morse or Liusternik-Schnirelman theories. The interested reader is referred to Ekeland [41], Mawhin-Willem [76], Struwe [97] or Zeidler [104].

$$f(x, u, \xi) = f(u, \xi) = 2\pi u \frac{\xi^3}{1 + \xi^2}$$

and

$$(P) \quad \inf \left\{ I(u) = \int_a^b f(u(x), u'(x)) \, dx : u(a) = \alpha, \ u(b) = \beta \right\} = m.$$

We will not treat this problem in the present book and we refer to Buttazzo-Kawohl [18] for a review article on this subject.

Example: brachistochrone. The aim is to find the shortest path between two points that follows a point mass moving under the influence of gravity. We place the initial point at the origin and the end one at $(b, -\beta)$, with $b, \beta > 0$. We let the gravity act downwards along the y-axis and we represent any point along the path by $(x, -u(x))$, $0 \leq x \leq b$.

In terms of our notations, we have that $n = N = 1$ and the function, under consideration, is

$$f(x, u, \xi) = f(u, \xi) = \frac{\sqrt{1 + \xi^2}}{\sqrt{2gu}}$$

and

$$(P) \quad \inf \left\{ I(u) = \int_0^b f(u(x), u'(x)) \, dx : u \in X \right\} = m$$

where

$$X = \left\{ u \in C^1([0, b]) : u(0) = 0, \ u(b) = \beta \text{ and } u(x) > 0, \ \forall x \in (0, b] \right\}.$$

The shortest path turns out to be a *cycloid*.

Example: minimal surface of revolution. We have to determine among all surfaces of revolution of the form

$$v(x, y) = (x, u(x) \cos y, u(x) \sin y)$$

with fixed end points $u(a) = \alpha$, $u(b) = \beta$ one with minimal area. We still have $n = N = 1$,

$$f(x, u, \xi) = f(u, \xi) = 2\pi u \sqrt{1 + \xi^2}$$

and

$$(P) \quad \inf \left\{ I(u) = \int_a^b f(u(x), u'(x)) \, dx : u(a) = \alpha, \ u(b) = \beta, \ u > 0 \right\} = m.$$

with $u : \overline{\Omega} \to \mathbb{R}$ and where $\Omega \subset \mathbb{R}^n$ is a bounded domain. These surfaces are therefore graphs of functions. The fact that $\partial\Sigma$ is a preassigned curve Γ, reads now as $u = u_0$ on $\partial\Omega$, where u_0 is a given function. The area of such a surface is given by

$$\text{Area}(\Sigma) = I(u) = \int_{\Omega} f(\nabla u(x)) \, dx$$

where, for $\xi \in \mathbb{R}^n$, we have set

$$f(\xi) = \sqrt{1 + |\xi|^2}.$$

The problem is then written in the usual form

$$(P) \quad \inf\left\{ I(u) = \int_{\Omega} f(\nabla u(x)) \, dx : u = u_0 \text{ on } \partial\Omega \right\}.$$

Associated with (P) we have the so-called *minimal surface equation*

$$(E) \quad Mu \equiv \left(1 + |\nabla u|^2\right) \Delta u - \sum_{i,j=1}^{n} u_{x_i} u_{x_j} u_{x_i x_j} = 0$$

which is the equation that any minimizer u of (P) should satisfy. In geometrical terms, this equation just expresses the fact that the corresponding surface Σ has everywhere vanishing *mean curvature*.

Case 2: parametric surfaces. Nonparametric surfaces are clearly too restrictive from the geometrical point of view and one is led to consider *parametric surfaces*. These are sets $\Sigma \subset \mathbb{R}^{n+1}$ so that there exist a domain $\Omega \subset \mathbb{R}^n$ and a map $v : \overline{\Omega} \to \mathbb{R}^{n+1}$ such that

$$\Sigma = v\left(\overline{\Omega}\right) = \left\{ v(x) : x \in \overline{\Omega} \right\}.$$

For example, when $n = 2$ and $v = v(x_1, x_2) \in \mathbb{R}^3$, if we denote by $v_{x_1} \times v_{x_2}$ the normal to the surface (where $a \times b$ stands for the vectorial product of $a, b \in \mathbb{R}^3$ and $v_{x_1} = \partial v / \partial x_1$, $v_{x_2} = \partial v / \partial x_2$) we find that the area is given by

$$\text{Area}(\Sigma) = J(v) = \iint_{\Omega} |v_{x_1} \times v_{x_2}| \, dx_1 dx_2.$$

In terms of the notations introduced at the beginning of the present section we have $n = 2$ and $N = 3$.

Example: isoperimetric inequality. Let $A \subset \mathbb{R}^2$ be a bounded open set whose boundary, ∂A, is a sufficiently regular simple closed curve. Denote by

Then, by analyzing the behavior of the higher derivatives of F, we determine if $\overline{x}$ is a minimum (global or local), a maximum (global or local) or just a stationary point.

The second method consists of considering a minimizing sequence $\{x_\nu\} \subset X$ so that

$$F(x_\nu) \to \inf\{F(x) : x \in X\}.$$

We then, with appropriate hypotheses on F, prove that the sequence is compact in X, meaning that

$$x_\nu \to \overline{x} \in X, \text{ as } \nu \to \infty.$$

Finally if F is lower semicontinuous, meaning that

$$\liminf_{\nu \to \infty} F(x_\nu) \geq F(\overline{x})$$

we have indeed shown that $\overline{x}$ is a minimizer of (P).

We can proceed in a similar manner for problems of the calculus of variations. The first and second methods are then called, respectively, classical and direct methods. However, the problem is now considerably harder because we are working in infinite dimensional spaces.

Let us recall the problem under consideration

$$(P) \quad \inf\left\{I(u) = \int_\Omega f(x, u(x), \nabla u(x)) \, dx : u \in X\right\} = m$$

where

- $\Omega \subset \mathbb{R}^n$, $n \geq 1$, is a bounded open set, points in Ω are denoted by $x = (x_1, \cdots, x_n)$;

- $u : \Omega \to \mathbb{R}^N$, $N \geq 1$, $u = (u^1, \cdots, u^N)$ and $\nabla u = \left(\frac{\partial u^j}{\partial x_i}\right)_{1 \leq i \leq n}^{1 \leq j \leq N} \in \mathbb{R}^{N \times n}$;

- $f : \overline{\Omega} \times \mathbb{R}^N \times \mathbb{R}^{N \times n} \longrightarrow \mathbb{R}$, $f = f(x, u, \xi)$, is continuous;

- X is a space of admissible functions which satisfy $u = u_0$ on $\partial\Omega$, where u_0 is a given function.

Here, contrary to the case of $\mathbb{R}^N$, we encounter a preliminary problem, namely: what is the best choice for the space X of admissible functions. A natural one seems to be $X = C^1(\overline{\Omega})$. There are several reasons, which will be clearer during the course of the book, that indicate that this is not the best choice. A better one is the *Sobolev space* $W^{1,p}(\Omega)$, $p \geq 1$. We say that $u \in W^{1,p}(\Omega)$, if u is (weakly) differentiable and if

$$\|u\|_{W^{1,p}} = \left[\int_\Omega (|u(x)|^p + |\nabla u(x)|^p) \, dx\right]^{1/p} < \infty$$

topology that we have on $W^{1,p}$. The natural one is the one induced by the norm, that we call *strong convergence* and that we denote by

$$u_\nu \to \overline{u} \text{ in } W^{1,p}.$$

However, it is, in general, not an easy matter to show that the sequence converges in such a strong topology. It is often better to weaken the notion of convergence and to consider the so-called *weak convergence*, denoted by $\rightharpoonup$. To obtain that

$$u_\nu \rightharpoonup \overline{u} \text{ in } W^{1,p}, \text{ as } \nu \to \infty$$

is much easier and it is enough, for example if $p > 1$, to show (up to the extraction of a subsequence) that

$$\|u_\nu\|_{W^{1,p}} \leq \gamma$$

where γ is a constant independent of ν. Such an estimate follows, for instance, if we impose a *coercivity* assumption on the function f of the type

$$\lim_{|\xi| \to \infty} \frac{f(x, u, \xi)}{|\xi|} = +\infty, \ \forall (x, u) \in \overline{\Omega} \times \mathbb{R}.$$

We observe that the Dirichlet integral, with

$$f(x, u, \xi) = \frac{1}{2} |\xi|^2,$$

satisfies this hypothesis but not the minimal surface in nonparametric form, where

$$f(x, u, \xi) = \sqrt{1 + |\xi|^2}.$$

The second step consists of showing that the functional I is lower semicontinuous with respect to weak convergence, namely

$$u_\nu \rightharpoonup \overline{u} \text{ in } W^{1,p} \ \Rightarrow \ \liminf_{\nu \to \infty} I(u_\nu) \geq I(\overline{u}).$$

We will see that this conclusion is true if

$$\xi \to f(x, u, \xi) \text{ is convex}, \forall (x, u) \in \overline{\Omega} \times \mathbb{R}.$$

Since $\{u_\nu\}$ was a minimizing sequence, we deduce that $\overline{u}$ is indeed a minimizer of (P).

In Chapter 5 we consider the problem of minimal surfaces. The methods of Chapter 3 cannot be directly applied. In fact the step of compactness of the minimizing sequences is much harder to obtain, for reasons that we explain in Chapter 5. There are, moreover, difficulties related to the geometrical nature of

Brézis [14] and Evans [44] for a very clear introduction to the subject. The monographs of Edmunds-Evans [40] and Gilbarg-Trudinger [51] can also be of great help. The book of Adams [1] is surely one of the most complete in this field, but its reading is harder than the other four.

Finally in Section 1.5 we gather some important properties of convex functions such as Jensen inequality, the Legendre transform and Carathéodory theorem. The book of Rockafellar [91] is classical in this field. One can also consult Hörmander [64] or Webster [101], see also [31].

1.2   Continuous and Hölder continuous functions

Definition 1.1 *Let $\Omega \subset \mathbb{R}^n$ be an open set.*

(i) $C^0(\Omega) = C(\Omega)$ is the set of continuous functions $u : \Omega \to \mathbb{R}$.

(ii) $C^0(\Omega; \mathbb{R}^N) = C(\Omega; \mathbb{R}^N)$ is the set of continuous maps $u : \Omega \to \mathbb{R}^N$, *meaning that if $u = (u^1, \cdots, u^N)$, then $u^i \in C(\Omega)$, for every $i = 1, \cdots, N$.*

(iii) $C^0(\overline{\Omega}) = C(\overline{\Omega})$ is the set of bounded continuous functions $u : \Omega \to \mathbb{R}$, which are extended in a continuous and bounded way to $\overline{\Omega}$.

(iv) When we are dealing with maps, $u : \Omega \to \mathbb{R}^N$, we write, similarly as above, $C^0(\overline{\Omega}; \mathbb{R}^N) = C(\overline{\Omega}; \mathbb{R}^N)$.

(v) The support *of a function $u : \Omega \to \mathbb{R}$ is defined as*

$$\operatorname{supp} u = \overline{\{x \in \Omega : u(x) \neq 0\}}.$$

(vi) $C_0(\Omega) = \{u \in C(\Omega) : \operatorname{supp} u \subset \Omega \text{ is compact}\}$.

(vii) We define the norm over $C(\overline{\Omega})$, by

$$\|u\|_{C^0} = \sup_{x \in \overline{\Omega}} |u(x)|.$$

Remark 1.2 (i) $C(\overline{\Omega})$ equipped with the norm $\|\cdot\|_{C^0}$ is a *Banach space*.

(ii) In the definition of $C(\overline{\Omega})$ we have required that the functions be bounded. If Ω is bounded, this is not a restriction. If Ω is unbounded, we could, as some authors do, require that the functions be only continuously extended to the boundary (and not necessarily bounded). In the framework of the present book the definition that we adopt is however more appropriate, since we want to define a norm on $C(\overline{\Omega})$ even if Ω is unbounded. Note that, according to our definition, $C(\overline{\mathbb{R}^n}) \subsetneq C(\mathbb{R}^n)$; indeed the function $u(x) = x$ is such that $u \in C(\mathbb{R})$, but $u \notin C(\overline{\mathbb{R}})$.

(iii) $C_0^k(\Omega) = C^k(\Omega) \cap C_0(\Omega)$.

(iv) $C^\infty(\Omega) = \bigcap_{k=0}^{\infty} C^k(\Omega)$, $C^\infty(\overline{\Omega}) = \bigcap_{k=0}^{\infty} C^k(\overline{\Omega})$.

(v) $C_0^\infty(\Omega) = \mathcal{D}(\Omega) = C^\infty(\Omega) \cap C_0(\Omega)$.

(vi) When dealing with maps $u : \Omega \to \mathbb{R}^N$, we write, for example, $C^k(\Omega; \mathbb{R}^N)$, and similarly for the other cases.

Remark 1.5 **(i)** $C^k(\overline{\Omega})$ with its norm $\|\cdot\|_{C^k}$ is a *Banach space.*

(ii) When Ω is unbounded, see Remark 1.2.

We also need to define the set of piecewise continuous functions.

Definition 1.6 *Let $\Omega \subset \mathbb{R}^n$ be an open set.*

(i) Define $C_{piec}^0(\overline{\Omega}) = C_{piec}(\overline{\Omega})$ to be the set of bounded piecewise continuous functions $u : \overline{\Omega} \to \mathbb{R}$. *This means that there exists a finite (or more generally a countable) partition of Ω into open sets $\Omega_i \subset \Omega$, $i = 1, \cdots, I$, so that*

$$\overline{\Omega} = \bigcup_{i=1}^{I} \overline{\Omega}_i, \quad \Omega_i \cap \Omega_j = \emptyset, \text{ if } i \neq j, \ 1 \leq i, j \leq I$$

and $u|_{\overline{\Omega}_i}$ is bounded and continuous.

(ii) Similarly $C_{piec}^k(\overline{\Omega})$, $k \geq 1$, is the set of functions $u \in C^{k-1}(\overline{\Omega})$, whose partial derivatives of order k are in $C_{piec}^0(\overline{\Omega})$.

We now turn to the notion of Hölder continuous functions (examples are given in exercises).

Definition 1.7 *Let $D \subset \mathbb{R}^n$, $u : D \to \mathbb{R}$ and $0 < \alpha \leq 1$. We let*

$$[u]_{C^{0,\alpha}(D)} = \sup_{\substack{x,y \in D \\ x \neq y}} \left\{ \frac{|u(x) - u(y)|}{|x - y|^\alpha} \right\}.$$

Let $\Omega \subset \mathbb{R}^n$ be an open set, $k \geq 0$ be an integer. We define the different spaces of Hölder continuous functions in the following way.

(i) $C^{0,\alpha}(\Omega)$ is the set of $u \in C(\Omega)$ so that

$$[u]_{C^{0,\alpha}(K)} = \sup_{\substack{x,y \in K \\ x \neq y}} \left\{ \frac{|u(x) - u(y)|}{|x - y|^\alpha} \right\} < \infty$$

for every compact set $K \subset \Omega$.

Proposition 1.9 *Let* $\Omega \subset \mathbb{R}^n$ *be bounded and open,* $0 \leq \alpha \leq \beta \leq 1$ *and* $k \geq 0$ *be an integer. The following properties then hold.*

(i) *If* $u, v \in C^{0,\alpha}\left(\overline{\Omega}\right)$ *then* $uv \in C^{0,\alpha}\left(\overline{\Omega}\right)$.

(ii) *The following set of inclusions is valid*

$$C^k\left(\overline{\Omega}\right) \supset C^{k,\alpha}\left(\overline{\Omega}\right) \supset C^{k,\beta}\left(\overline{\Omega}\right) \supset C^{k,1}\left(\overline{\Omega}\right).$$

(iii) *If, in addition,* Ω *is convex, then*

$$C^{k,1}\left(\overline{\Omega}\right) \supset C^{k+1}\left(\overline{\Omega}\right).$$

Remark 1.10 (i) The proposition is proved in Exercise 1.2.2 below when $k = 0$.

(ii) The statement (i) extends to higher derivatives. More precisely if $u, v \in C^{k,\alpha}\left(\overline{\Omega}\right)$ then $uv \in C^{k,\alpha}\left(\overline{\Omega}\right)$, provided the domain Ω is not too wild, for example if it has Lipschitz boundary (cf. Definition 1.41 below for the precise meaning).

(iii) The inclusion in (iii) remains valid for very general domains Ω, for example those with Lipschitz boundary; however it is false in general, cf. Exercise 1.2.3.

1.2.1 Exercises

Exercise 1.2.1 (i) Let $\Omega = (0,1)$ and $u_\alpha(x) = x^\alpha$ with $\alpha \in (0,1]$. Show that $u_\alpha \in C^{0,\alpha}([0,1])$.

(ii) Prove that

$$u(x) = \begin{cases} -1/\log x & \text{if } x > 0 \\ 0 & \text{if } x = 0 \end{cases}$$

is continuous but not in $C^{0,\alpha}([0,1/2])$ for any $\alpha \in (0,1]$.

(iii) Let $\lambda \in (0,1]$ and

$$u_\lambda(x) = \sum_{n=1}^{\infty} \frac{\cos(2^n x)}{2^{n\lambda}}.$$

Show that $u_\lambda \in C^{0,\alpha}([0,\pi])$, for any $0 < \alpha < \lambda$.

Exercise 1.2.2 Show Proposition 1.9 when $k = 0$.

Exercise 1.2.3 Let $\frac{1}{2} < \beta < 1$,

$$\Omega = \left\{ (x_1, x_2) \in \mathbb{R}^2 : x_2 < \sqrt{|x_1|} \quad \text{and} \quad x_1^2 + x_2^2 < 1 \right\}$$

Remark 1.12 **(i)** The abbreviation "a.e." means that a property holds almost everywhere. For example, the function

$$\chi_{\mathbb{Q}}(x) = \begin{cases} 1 & \text{if } x \in \mathbb{Q} \\ 0 & \text{if } x \notin \mathbb{Q} \end{cases}$$

where $\mathbb{Q}$ is the set of rational numbers, is such that $\chi_{\mathbb{Q}} = 0$ a.e.

(ii) We adopt the convention that we identify two functions that coincide except on a set of measure zero. So, strictly speaking, the spaces L^p are equivalence classes.

In the next theorem we summarize the most important properties of L^p spaces that we need. We however do not recall Fatou lemma, the dominated convergence theorem and other basic theorems of Lebesgue integral.

Theorem 1.13 *Let $\Omega \subset \mathbb{R}^n$ be open and $1 \leq p \leq \infty$.*

(i) $\|\cdot\|_{L^p}$ is a norm and $L^p(\Omega)$, equipped with this norm, is a Banach space. The space $L^2(\Omega)$ is a Hilbert space with scalar product given by

$$\langle u; v \rangle = \int_{\Omega} u(x) v(x) \, dx.$$

*(ii) **Hölder inequality** asserts that if $u \in L^p(\Omega)$ and $v \in L^{p'}(\Omega)$ where $1/p + 1/p' = 1$ (i.e., $p' = p/(p-1)$) and $1 \leq p \leq \infty$ then $uv \in L^1(\Omega)$ and moreover*

$$\|uv\|_{L^1} \leq \|u\|_{L^p} \|v\|_{L^{p'}}.$$

*(iii) **Minkowski inequality** asserts that*

$$\|u + v\|_{L^p} \leq \|u\|_{L^p} + \|v\|_{L^p}.$$

*(iv) **Riesz theorem**: the dual space of L^p, denoted by $(L^p)'$, can be identified with $L^{p'}(\Omega)$ where $1/p + 1/p' = 1$ provided $1 \leq p < \infty$. More precisely if $\varphi \in (L^p)'$ with $1 \leq p < \infty$ then there exists a unique $u \in L^{p'}$ so that*

$$\langle \varphi; f \rangle = \varphi(f) = \int_{\Omega} u(x) f(x) \, dx, \ \forall f \in L^p(\Omega)$$

and moreover

$$\|u\|_{L^{p'}} = \|\varphi\|_{(L^p)'}.$$

(v) L^p is separable if $1 \leq p < \infty$ and reflexive if $1 < p < \infty$.

(vi) Let $1 \leq p < \infty$. The piecewise constant functions (also called step functions if $\Omega \subset \mathbb{R}$), or the $C_0^\infty(\Omega)$ functions are dense in L^p. More precisely if $u \in L^p(\Omega)$ then there exist $u_\nu \in C_0^\infty(\Omega)$ (or u_ν piecewise constants) so that

$$\lim_{\nu \to \infty} \|u_\nu - u\|_{L^p} = 0.$$

Remark 1.16 **(i)** We speak of weak $*$ convergence in L^∞ instead of weak convergence, because as seen above the dual of L^∞ is strictly larger than L^1. Formally, however, weak convergence in L^p and weak $*$ convergence in L^∞ take the same form.

(ii) The limit (weak or strong) is unique.

(iii) It is obvious that

$$u_\nu \to u \text{ in } L^p \ \Rightarrow \ \begin{cases} u_\nu \rightharpoonup u \text{ in } L^p & \text{if } 1 \le p < \infty \\ u_\nu \overset{*}{\rightharpoonup} u \text{ in } L^\infty & \text{if } p = \infty. \end{cases}$$

Example 1.17 (cf. Exercise 1.3.2). Let $\Omega = (0,1)$, $\alpha \ge 0$ and

$$u_\nu(x) = \begin{cases} \nu^\alpha & \text{if } x \in (0, 1/\nu) \\ 0 & \text{if } x \in (1/\nu, 1). \end{cases}$$

If $1 < p < \infty$, we find

$$u_\nu \to 0 \text{ in } L^p \ \Leftrightarrow \ 0 \le \alpha < 1/p$$
$$u_\nu \rightharpoonup 0 \text{ in } L^p \ \Leftrightarrow \ 0 \le \alpha \le 1/p.$$

Example 1.18 Let $\Omega = (0, 2\pi)$ and $u_\nu(x) = \sin \nu x$, then

$$u_\nu \nrightarrow 0 \text{ in } L^p, \ \forall 1 \le p \le \infty$$
$$u_\nu \rightharpoonup 0 \text{ in } L^p, \ \forall 1 \le p < \infty$$

and

$$u_\nu \overset{*}{\rightharpoonup} 0 \text{ in } L^\infty.$$

These facts are consequences of Riemann-Lebesgue theorem (cf. Theorem 1.22).

Example 1.19 Let $\Omega = (0,1)$, $\alpha, \beta \in \mathbb{R}$

$$u(x) = \begin{cases} \alpha & \text{if } x \in (0, 1/2) \\ \beta & \text{if } x \in (1/2, 1). \end{cases}$$

Extend u by periodicity from $(0,1)$ to $\mathbb{R}$ and define

$$u_\nu(x) = u(\nu x).$$

Note that u_ν takes only the values α and β and the sets where it takes such values are, both, sets of measure $1/2$. It is clear that $\{u_\nu\}$ cannot be compact in any L^p spaces; however from Riemann-Lebesgue theorem (cf. Theorem 1.22), we find

$$u_\nu \rightharpoonup \frac{\alpha + \beta}{2} \text{ in } L^p, \ \forall 1 \le p < \infty \quad \text{and} \quad u_\nu \overset{*}{\rightharpoonup} \frac{\alpha + \beta}{2} \text{ in } L^\infty.$$

Theorem 1.22 (Riemann-Lebesgue theorem) *Let $1 \leq p \leq \infty$ and $u \in L^p(\Omega)$ where $\Omega = \prod_{i=1}^n (a_i, b_i)$. Let u be extended by periodicity from Ω to $\mathbb{R}^n$ and define*

$$u_\nu(x) = u(\nu x) \quad and \quad \overline{u} = \frac{1}{\operatorname{meas} \Omega} \int_\Omega u(x) \, dx$$

then $u_\nu \rightharpoonup \overline{u}$ in L^p if $1 \leq p < \infty$ and, if $p = \infty$, $u_\nu \overset{*}{\rightharpoonup} \overline{u}$ in L^∞.

Proof. To make the argument simpler we assume in the proof that $n = 1$, $\Omega = (0,1)$ and $1 < p \leq \infty$. For the proof of the general case ($\Omega \subset \mathbb{R}^n$ or $p = 1$) see, for example, Theorem 2.1.5 in [31]. We also assume, without loss of generality, that

$$\overline{u} = \int_0^1 u(x) \, dx = 0.$$

Step 1. Observe that if $1 \leq p < \infty$, then

$$\|u_\nu\|_{L^p}^p = \int_0^1 |u_\nu(x)|^p \, dx = \int_0^1 |u(\nu x)|^p \, dx$$

$$= \frac{1}{\nu} \int_0^\nu |u(y)|^p \, dy = \int_0^1 |u(y)|^p \, dy.$$

The last identity being a consequence of the 1-periodicity of u. We therefore find that

$$\|u_\nu\|_{L^p} = \|u\|_{L^p} . \tag{1.1}$$

The result is trivially true if $p = \infty$.

Step 2. (For a slightly different proof of this step see Exercise 1.3.5.) We therefore have that $u_\nu \in L^p$ and, since $\overline{u} = 0$, we have to show that

$$\lim_{\nu \to \infty} \int_0^1 u_\nu(x) \varphi(x) \, dx = 0, \ \forall \varphi \in L^{p'}(0,1) . \tag{1.2}$$

Let $\epsilon > 0$ be arbitrary. Since $\varphi \in L^{p'}(0,1)$ and $1 < p \leq \infty$, which implies $1 \leq p' < \infty$ (i.e. $p' \neq \infty$), we have from Theorem 1.13 that there exists h a step function so that

$$\|\varphi - h\|_{L^{p'}} \leq \epsilon . \tag{1.3}$$

Since h is a step function, we can find $a_0 = 0 < a_1 < \cdots < a_I = 1$ and $\alpha_i \in \mathbb{R}$ such that

$$h(x) = \alpha_i \text{ if } x \in (a_{i-1}, a_i), \ 1 \leq i \leq I.$$

We now write

$$\int_0^1 u_\nu(x) \varphi(x) \, dx = \int_0^1 u_\nu(x) [\varphi(x) - h(x)] \, dx + \int_0^1 u_\nu(x) h(x) \, dx$$

Definition 1.23 *Let $\Omega \subset \mathbb{R}^n$ be an open set and $1 \leq p \leq \infty$. We say that $u \in L^p_{loc}(\Omega)$ if $u \in L^p(\Omega')$ for every open set Ω' compactly contained in Ω (i.e. $\overline{\Omega'} \subset \Omega$ and $\overline{\Omega'}$ is compact).*

Theorem 1.24 (Fundamental lemma of the calculus of variations) *Let Ω be an open set of $\mathbb{R}^n$ and $u \in L^1_{loc}(\Omega)$ be such that*

$$\int_\Omega u(x)\,\psi(x)\,dx = 0, \ \forall \psi \in C_0^\infty(\Omega) \tag{1.5}$$

then $u = 0$, almost everywhere in Ω.

Proof. We prove the theorem under the stronger hypothesis that $u \in L^2(\Omega)$ and not only $u \in L^1_{loc}(\Omega)$ (recall that $L^2(\Omega) \subset L^1_{loc}(\Omega)$); for a proof in the general framework see, for example, Corollary 3.26 in Adams [1] or Lemma IV.2 in Brézis [14]. Let $\epsilon > 0$. Since $u \in L^2(\Omega)$, invoking Theorem 1.13, we can find $\psi \in C_0^\infty(\Omega)$ so that

$$\|u - \psi\|_{L^2} \leq \epsilon.$$

Using (1.5) we deduce that

$$\|u\|_{L^2}^2 = \int_\Omega u^2\,dx = \int_\Omega u(u - \psi)\,dx.$$

Combining the above identity and Hölder inequality, we find

$$\|u\|_{L^2}^2 \leq \|u\|_{L^2}\,\|u - \psi\|_{L^2} \leq \epsilon\,\|u\|_{L^2}\ .$$

Since $\epsilon > 0$ is arbitrary, we deduce that $\|u\|_{L^2} = 0$ and hence the claim. ∎

We next have as a consequence the following result (for a proof see Exercise 1.3.6 and for a generalization see Exercise 1.3.7).

Corollary 1.25 *Let $\Omega \subset \mathbb{R}^n$ be an open set and $u \in L^1_{loc}(\Omega)$ be such that*

$$\int_\Omega u(x)\,\psi(x)\,dx = 0, \ \forall \psi \in C_0^\infty(\Omega) \ \text{with} \ \int_\Omega \psi(x)\,dx = 0$$

then $u = constant$, almost everywhere in Ω.

1.3.1   Exercises

Exercise 1.3.1 (i) Prove Hölder and Minkowski inequalities.

    (ii) Show that if $p, q \geq 1$ with $pq/(p+q) \geq 1$, $u \in L^p$ and $v \in L^q$, then

$$uv \in L^{pq/p+q} \quad \text{and} \quad \|uv\|_{L^{pq/p+q}} \leq \|u\|_{L^p}\,\|v\|_{L^q}\ .$$

Exercise 1.3.5 In Step 2 of Theorem 1.22 use approximation by smooth functions instead of by step functions and anticipate on (1.11) and (1.12) in Lemma 1.39.

Exercise 1.3.6 (i) Show Corollary 1.25.
   (ii) Prove that if $u \in L^1_{loc}(a,b)$ is such that

$$\int_a^b u(x)\, \varphi'(x)\, dx = 0, \ \forall \varphi \in C_0^\infty(a,b)$$

then $u = $ constant, almost everywhere in (a,b).

Exercise 1.3.7 Generalize Corollary 1.25 in the following manner. Let $\Omega \subset \mathbb{R}^n$ be an open set and $\alpha_1, \cdots, \alpha_N \in L^1_{loc}(\Omega)$. Let

$$X = \left\{ \psi \in C_0^\infty(\Omega) : \int_\Omega \alpha_i(x)\, \psi(x)\, dx = 0, \ i = 1, \cdots, N \right\}$$

and $u \in L^1_{loc}(\Omega)$ be such that

$$\int_\Omega u(x)\, \psi(x)\, dx = 0, \ \forall \psi \in X.$$

Show that there exist constants $a_1, \cdots, a_N \in \mathbb{R}$ such that

$$u(x) = \sum_{i=1}^N a_i\, \alpha_i(x) \quad \text{a.e. } x \in \Omega.$$

Suggestion: use the elementary algebraic result (see Lemma 3.9 page 62 in [93]) which asserts that if X is a vector space and Λ, Λ_i, $i = 1, \cdots, N$, are linear functionals on X such that

$$\Lambda(x) = 0, \text{ for every } x \in X \text{ with } \Lambda_i(x) = 0, \ i = 1, \cdots, N;$$

then there exist constants $a_1, \cdots, a_N \in \mathbb{R}$ such that

$$\Lambda = \sum_{i=1}^N a_i\, \Lambda_i\,.$$

Exercise 1.3.8 Generalize Exercise 1.3.6 (ii) in the following way. Prove that if $n \geq 1$ is an integer and $u \in L^1_{loc}(a,b)$ is such that

$$\int_a^b u(x)\, \varphi^{(n)}(x)\, dx = 0, \ \forall \varphi \in C_0^\infty(a,b)$$

differentiated in this way. In particular a discontinuous function of $\mathbb{R}$ cannot be differentiated in the weak sense (see Example 1.29).

Example 1.28 Let $\Omega=\mathbb{R}$ and the function $u(x) = |x|$. Its weak derivative is then given by

$$u'(x) = \begin{cases} +1 & \text{if } x > 0 \\ -1 & \text{if } x < 0. \end{cases}$$

Example 1.29 (Dirac mass) Let

$$H(x) = \begin{cases} +1 & \text{if } x > 0 \\ 0 & \text{if } x < 0. \end{cases}$$

We now show that H has no weak derivative. Let $\Omega = (-1,1)$. Assume, for the sake of contradiction, that $H' = \delta \in L^1_{\text{loc}}(-1,1)$ and let us prove that this is absurd. Let $\varphi \in C_0^\infty(0,1)$ be arbitrary and extend it to $(-1,0)$ by $\varphi \equiv 0$. We therefore have by definition that

$$\int_{-1}^1 \delta(x)\,\varphi(x)\,dx = -\int_{-1}^1 H(x)\,\varphi'(x)\,dx = -\int_0^1 \varphi'(x)\,dx$$
$$= \varphi(0) - \varphi(1) = 0.$$

We hence find

$$\int_0^1 \delta(x)\,\varphi(x)\,dx = 0, \ \forall \varphi \in C_0^\infty(0,1)$$

which combined with Theorem 1.24, leads to $\delta = 0$ a.e. in $(0,1)$. With an analogous reasoning we would get that $\delta = 0$ a.e. in $(-1,0)$ and consequently $\delta = 0$ a.e. in $(-1,1)$. Let us show that we already reached the desired contradiction. Indeed if this were the case we would have, for every $\varphi \in C_0^\infty(-1,1)$,

$$0 = \int_{-1}^1 \delta(x)\,\varphi(x)\,dx = -\int_{-1}^1 H(x)\,\varphi'(x)\,dx$$
$$= -\int_0^1 \varphi'(x)\,dx = \varphi(0) - \varphi(1) = \varphi(0).$$

This would imply that $\varphi(0) = 0$, for every $\varphi \in C_0^\infty(-1,1)$, which is clearly absurd. Thus H is not weakly differentiable.

Remark 1.30 By weakening even more the notion of derivative (for example, by not requiring anymore that v is in L^1_{loc}), the theory of distributions can give a meaning at $H' = \delta$, it is then called the *Dirac mass*. We will however not need this theory in the sequel, except, but only marginally, in the exercises of Section 3.5.

Example 1.33 The following cases are discussed in Exercise 1.4.1.

(i) Let $s > 0$,

$$\Omega = \{x \in \mathbb{R}^n : |x| < 1\} \quad \text{and} \quad \psi(x) = |x|^{-s}.$$

We then have

$$\psi \in L^p \Leftrightarrow sp < n \quad \text{and} \quad \psi \in W^{1,p} \Leftrightarrow (s+1)p < n.$$

(ii) Let $0 < s < 1/2$,

$$\Omega = \{x = (x_1, x_2) \in \mathbb{R}^2 : |x| < 1/2\} \quad \text{and} \quad \psi(x) = |\log |x||^s.$$

We have that $\psi \in W^{1,2}(\Omega)$, $\psi \in L^p(\Omega)$ for every $1 \le p < \infty$, but $\psi \notin L^\infty(\Omega)$.

(iii) Let

$$\Omega = \{x \in \mathbb{R}^n : |x| < 1\} \quad \text{and} \quad u(x) = \frac{x}{|x|},$$

then $u \in W^{1,p}(\Omega; \mathbb{R}^n)$ for every $1 \le p < n$.

Theorem 1.34 *Let $\Omega \subset \mathbb{R}^n$ be open, $1 \le p \le \infty$ and $k \ge 1$ an integer.*

(i) $W^{k,p}(\Omega)$ equipped with its norm $\|\cdot\|_{k,p}$ is a Banach space which is separable if $1 \le p < \infty$ and reflexive if $1 < p < \infty$.

(ii) $W^{1,2}(\Omega)$ is a Hilbert space when endowed with the following scalar product

$$\langle u; v \rangle_{W^{1,2}} = \int_\Omega u(x) v(x) \, dx + \int_\Omega \langle \nabla u(x); \nabla v(x) \rangle \, dx.$$

(iii) The $C^\infty(\Omega) \cap W^{k,p}(\Omega)$ functions are dense in $W^{k,p}(\Omega)$ provided $1 \le p < \infty$. Moreover, if Ω is a bounded domain with Lipschitz boundary (cf. Definition 1.41), then $C^\infty(\overline{\Omega})$ is also dense in $W^{k,p}(\Omega)$ provided $1 \le p < \infty$.

(iv) $W_0^{k,p}(\mathbb{R}^n) = W^{k,p}(\mathbb{R}^n)$, whenever $1 \le p < \infty$.

Remark 1.35 (i) Note that as for the case of L^p the space $W^{k,p}$ is reflexive only when $1 < p < \infty$ and hence $W^{1,1}$ is not reflexive; as already said, this is the main source of difficulties in the minimal surface problem.

(ii) The density result is due to Meyers and Serrin, see Theorem 3.16 in Adams [1], Section 5.3 in Evans [44] or Section 7.6 in Gilbarg-Trudinger [51].

(iii) In general, we have $W_0^{1,p}(\Omega) \subsetneq W^{1,p}(\Omega)$, but when $\Omega = \mathbb{R}^n$ both coincide (see Corollary 3.19 in Adams [1]).

Proof. We prove the theorem only when $n = 1$ and $\Omega = (a, b)$. For the more general case see, for example, Proposition IX.3 in Brézis [14] or Theorem 5.8.3 and 5.8.4 in Evans [44].

(i) $\Rightarrow$ **(ii)**. This follows from Hölder inequality and the fact that u has a weak derivative; indeed

$$\left| \int_a^b u\,(x)\,\varphi'\,(x)\; dx \right| = \left| \int_a^b u'\,(x)\,\varphi\,(x)\; dx \right| \leq \|u'\|_{L^p}\,\|\varphi\|_{L^{p'}}\,.$$

(ii) $\Rightarrow$ **(i)**. Let F be a linear functional defined by

$$F\,(\varphi) = \langle F; \varphi \rangle = \int_a^b u\,(x)\,\varphi'\,(x)\; dx, \;\; \forall \varphi \in C_0^\infty\,(a, b)\,. \tag{1.6}$$

Note that, by (ii), it is continuous over $C_0^\infty\,(a, b)$. Since $C_0^\infty\,(a, b)$ is dense in $L^{p'}\,(a, b)$ (note that we used here the fact that $p \neq 1$ and hence $p' \neq \infty$), we can extend it, by continuity (or appealing to Hahn-Banach theorem), to the whole $L^{p'}\,(a, b)$; we have therefore defined a continuous linear operator F over $L^{p'}\,(a, b)$, with

$$|F\,(\varphi)| \leq \gamma\,\|\varphi\|_{L^{p'}}\,, \;\; \forall \varphi \in L^{p'}\,(a, b)\,. \tag{1.7}$$

From Riesz theorem (Theorem 1.13) we find that there exists $v \in L^p\,(a, b)$ so that

$$F\,(\varphi) = \langle F; \varphi \rangle = \int_a^b v\,(x)\,\varphi\,(x)\; dx, \;\; \forall \varphi \in L^{p'}\,(a, b)\,. \tag{1.8}$$

Combining (1.6) and (1.8) we get

$$\int_a^b (-v\,(x))\,\varphi\,(x)\; dx = - \int_a^b u\,(x)\,\varphi'\,(x)\; dx, \;\; \forall \varphi \in C_0^\infty\,(a, b)$$

which exactly means that $u' = -v \in L^p\,(a, b)$ and hence $u \in W^{1,p}\,(a, b)$.

Note also that, since (1.7) and (1.8) hold, we infer

$$\|u'\|_{L^p(a,b)} = \|v\|_{L^p(a,b)} \leq \gamma.$$

(iii) $\Rightarrow$ **(ii)**. Let $\varphi \in C_0^\infty\,(a, b)$ and let $\omega \subset \overline{\omega} \subset (a, b)$ with $\overline{\omega}$ compact and such that $\operatorname{supp} \varphi \subset \omega$. Let $\tau \in \mathbb{R}$ so that $0 \neq |\tau| < \operatorname{dist}\,(\omega, (a, b)^c)$. We then have for $1 < p \leq \infty$, appealing to Hölder inequality and to (iii),

$$\left| \int_a^b [D_\tau u\,(x)]\,\varphi\,(x)\; dx \right| \leq \|D_\tau u\|_{L^p(\omega)}\,\|\varphi\|_{L^{p'}(a,b)} \leq \gamma\,\|\varphi\|_{L^{p'}(a,b)}\,. \tag{1.9}$$

and hence
$$\|D_\tau u\|_{L^p(\omega)} \le \|u'\|_{L^p(a,b)}$$
which is the claim. ∎

In the proof of Theorem 1.37, we have used a result that, roughly speaking, says that functions in $W^{1,p}$ are continuous and are primitives of functions in L^p.

Lemma 1.39 *Let* $u \in W^{1,p}(a,b)$, $1 \le p \le \infty$. *Then there exists a function* $\widetilde{u} \in C([a,b])$ *such that* $u = \widetilde{u}$ *a.e. and*

$$\widetilde{u}(x) - \widetilde{u}(y) = \int_y^x u'(t)\, dt, \ \forall x, y \in [a,b].$$

Remark 1.40 (i) As already repeated, we ignore the difference between u and $\widetilde{u}$ and we say that if $u \in W^{1,p}(a,b)$ then $u \in C([a,b])$ and u is the primitive of u', i.e.

$$u(x) - u(y) = \int_y^x u'(t)\, dt.$$

(ii) Lemma 1.39 is a particular case of Sobolev imbedding theorem (cf. below). It gives a non-trivial result, in the sense that it is not, a priori, obvious that a function $u \in W^{1,p}(a,b)$ is continuous. We can therefore say that

$$C^1([a,b]) \subset W^{1,p}(a,b) \subset C([a,b]),\ 1 \le p \le \infty.$$

(iii) The inequality (1.13) in the proof of the lemma below shows that if $u \in W^{1,p}(a,b)$, $1 < p < \infty$, then $u \in C^{0,1/p'}([a,b])$ and hence u is Hölder continuous with exponent $1/p'$. We have already seen in Remark 1.38 that if $p = \infty$, then $C^{0,1}([a,b])$ and $W^{1,\infty}(a,b)$ can be identified.

Proof. We divide the proof into two steps.

Step 1. Let $c \in (a,b)$ be fixed and define

$$v(x) = \int_c^x u'(t)\, dt,\ x \in [a,b]. \tag{1.11}$$

Let us show that $v \in C([a,b])$ and

$$\int_a^b v(x)\, \varphi'(x)\, dx = -\int_a^b u'(x)\, \varphi(x)\, dx,\ \forall \varphi \in C_0^\infty(a,b). \tag{1.12}$$

Indeed we have

$$\int_a^b v(x)\, \varphi'(x)\, dx = \int_a^b \left(\int_c^x u'(t)\, dt \right) \varphi'(x)\, dx$$

$$= \int_a^c dx \int_c^x u'(t)\, \varphi'(x)\, dt + \int_c^b dx \int_c^x u'(t)\, \varphi'(x)\, dt$$

Definition 1.41 *(i)* *Let $\Omega \subset \mathbb{R}^n$ be open and bounded. We say that Ω is a bounded open set with C^k, $k \geq 1$, boundary if for every $x \in \partial\Omega$, there exist a neighborhood $U \subset \mathbb{R}^n$ of x and a one-to-one and onto map $H : Q \to U$, where*

$$Q = \{x \in \mathbb{R}^n : |x_j| < 1, \ j = 1, 2, \cdots, n\}$$

$$H \in C^k\left(\overline{Q}\right), \ H^{-1} \in C^k\left(\overline{U}\right), \ H\left(Q_+\right) = U \cap \Omega, \ H\left(Q_0\right) = U \cap \partial\Omega$$

with $Q_+ = \{x \in Q : x_n > 0\}$ and $Q_0 = \{x \in Q : x_n = 0\}$.

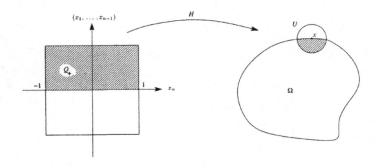

Figure 1.1: regular boundary

(ii) *If H and H^{-1} are in $C^{k,\alpha}$, $0 < \alpha \leq 1$, we say that Ω is a* bounded open set with $C^{k,\alpha}$ *boundary.*

(iii) *If H and H^{-1} are only in $C^{0,1}$, we say that Ω is a* bounded open set with Lipschitz *boundary.*

Remark 1.42 A polyhedron or a convex set has Lipschitz boundary, while the unit ball in $\mathbb{R}^n$ has a C^∞ boundary.

In the next two theorems (see for references Theorems 5.4 and 6.2 in Adams [1]) we write some inclusions between spaces; they have to be understood up to a choice of a representative.

Theorem 1.43 (Sobolev imbedding theorem) *Let $\Omega \subset \mathbb{R}^n$ be a bounded open set with Lipschitz boundary.*

Case 1. If $1 \leq p < n$ then

$$W^{1,p}\left(\Omega\right) \subset L^q\left(\Omega\right)$$

for every $p \geq 1$ (hence even when $p = 1$ we have that functions in $W^{1,1}$ are continuous). However, the imbedding is compact only when $p > 1$.

(ii) The result of case 1 in Theorem 1.44 is false if $q = p^*$.

(iii) In higher dimension, $n \geq 2$, the case $p = n$ in Theorem 1.43 cannot be improved, in general. The functions in $W^{1,n}$ are in general not continuous and not even bounded (cf. Example 1.33).

(iv) If Ω is unbounded, for example $\Omega = \mathbb{R}^n$, we must be more careful, in particular, the compactness of the imbeddings is lost (see the bibliography for more details).

(v) If we consider $W_0^{1,p}$ instead of $W^{1,p}$ then the same imbeddings are valid, but no restriction on the regularity of $\partial\Omega$ is required.

(vi) Similar imbeddings can be obtained if we replace $W^{1,p}$ by $W^{k,p}$.

(vii) Recall that $W^{1,\infty}(\Omega)$, when Ω is bounded and convex, is identified with $C^{0,1}(\overline{\Omega})$.

(viii) We now try to summarize the results when $n = 1$. If we denote by $I = (a, b)$, we have, for $p \geq 1$,

$$\mathcal{D}(I) = C_0^\infty(I) \subset \cdots \subset W^{2,p}(I) \subset C^1(\overline{I}) \subset W^{1,p}(I)$$
$$\subset C(\overline{I}) \subset L^\infty(I) \subset \cdots \subset L^2(I) \subset L^1(I)$$

and furthermore C_0^∞ is dense in L^1, equipped with its norm.

Theorems 1.43 and 1.44 will not be proved; they have been discussed in the one dimensional case in Lemma 1.39. Concerning the compactness of the imbedding when $n = 1$, it is a consequence of Ascoli-Arzelà theorem (see Exercise 1.4.4 for more details).

Before proceeding further it is important to understand the significance of Theorem 1.44. We are going to formulate it for sequences, since it is in this framework that we use it. The corollary says that if a sequence converges weakly in $W^{1,p}$, it, in fact, converges strongly in L^p.

Corollary 1.46 *Let $\Omega \subset \mathbb{R}^n$ be a bounded open set with Lipschitz boundary and $1 \leq p < \infty$. If*

$$u_\nu \rightharpoonup u \ in \ W^{1,p}(\Omega)$$

(this means that $u, u_\nu \in W^{1,p}(\Omega)$, $u_\nu \rightharpoonup u$ in L^p and $\nabla u_\nu \rightharpoonup \nabla u$ in L^p). Then

$$u_\nu \rightarrow u \ in \ L^p(\Omega).$$

If $p = \infty$, $u_\nu \stackrel{*}{\rightharpoonup} u$ in $W^{1,\infty}$, then $u_\nu \rightarrow u$ in L^∞.

We now prove that, for every $1 \leq p \leq \infty$,

$$\|u\|_{L^p} \leq (b-a) \|u'\|_{L^p} . \tag{1.14}$$

Since $u(a) = 0$, we have

$$|u(x)| = |u(x) - u(a)| = \left| \int_a^x u'(t) \, dt \right| \leq \int_a^b |u'(t)| \, dt = \|u'\|_{L^1} .$$

From this inequality we immediately get that (1.14) is true for $p = \infty$. When $p = 1$, we have after integration that

$$\|u\|_{L^1} = \int_a^b |u(x)| \, dx \leq (b-a) \|u'\|_{L^1} .$$

So it remains to prove (1.14) when $1 < p < \infty$. Applying Hölder inequality, we obtain

$$|u(x)| \leq \left(\int_a^b 1^{p'} \right)^{1/p'} \left(\int_a^b |u'|^p \right)^{1/p} = (b-a)^{1/p'} \|u'\|_{L^p}$$

and hence

$$\|u\|_{L^p} = \left(\int_a^b |u|^p \, dx \right)^{1/p}$$

$$\leq \left((b-a)^{\frac{p}{p'}} \|u'\|_{L^p}^p \int_a^b dx \right)^{1/p} = (b-a) \|u'\|_{L^p} .$$

This concludes the proof of the theorem when $n = 1$. ∎

1.4.1 Exercises

Exercise 1.4.1 Let $1 \leq p < \infty$, $R > 0$ and $B_R = \{x \in \mathbb{R}^n : |x| < R\}$. Let for $f \in C^\infty(0, +\infty)$ and for $x \in B_R$

$$u(x) = f(|x|) .$$

(i) Show that $u \in L^p(B_R)$ if and only if

$$\int_0^R r^{n-1} |f(r)|^p \, dr < \infty.$$

(ii) Assume that

$$\lim_{r \to 0} \left[r^{n-1} |f(r)| \right] = 0.$$

Exercise 1.4.6 Let $1 < p < \infty$, $u, u_\nu \in W^{1,p}(\Omega)$ and $\gamma > 0$ be such that

$$u_\nu \rightharpoonup u \text{ in } L^p \quad \text{and} \quad \|u\|_{W^{1,p}}, \|u_\nu\|_{W^{1,p}} \le \gamma.$$

Prove that

$$u_\nu \rightharpoonup u \text{ in } W^{1,p}$$

and that the result is also valid if $p = \infty$ and weak convergence is replaced by weak $*$ convergence.

Exercise 1.4.7 Let $\Omega = (0,1) \times (0,1) \subset \mathbb{R}^2$ and

$$u_\nu(x_1, x_2) = \frac{1}{\sqrt{\nu}}(1 - x_2)^\nu \sin \nu x_1.$$

Prove that $u_\nu \to 0$ in L^∞, $\|\nabla u_\nu\|_{L^2} \le \gamma$ for some constant $\gamma > 0$ and

$$u_\nu \rightharpoonup 0 \text{ in } W^{1,2}.$$

Exercise 1.4.8 Let $u \in W^{1,p}(\Omega)$ and $\varphi \in W_0^{1,p'}(\Omega)$, where $1/p + 1/p' = 1$ and $p > 1$. Show that

$$\int_\Omega u_{x_i}\varphi \, dx = -\int_\Omega u\varphi_{x_i} \, dx, \ i = 1, \cdots, n.$$

Exercise 1.4.9 Let $1 \le p < \infty$. Prove Poincaré inequality in the following way.

(i) Show that it is sufficient to establish the inequality for functions in $C_0^\infty(\Omega)$.

(ii) Let $Q = (-R, R)^n$ and $u \in C^1(\overline{Q})$ be such that

$$u(-R, x_2, \cdots, x_n) = 0, \text{ for every } -R \le x_2, \cdots, x_n \le R.$$

Prove (using Jensen inequality, cf. Theorem 1.52) that

$$\|u\|_{L^p} \le (2R)\|\nabla u\|_{L^p}.$$

(iii) Conclude.

Exercise 1.4.10 Let $\Omega \subset \mathbb{R}^n$ be a bounded open set, $1 < p < \infty$ and $u \in W_0^{1,p}(\Omega)$. Show that

$$\tilde{u}(x) = \begin{cases} u(x) & \text{if } x \in \Omega \\ 0 & \text{if } x \notin \Omega \end{cases}$$

is in $W^{1,p}(\mathbb{R}^n)$.

The following inequality is important (and is proved in a particular case in Exercise 1.5.2).

Theorem 1.52 (Jensen inequality) *Let $\Omega \subset \mathbb{R}^n$ be open and bounded, $u \in L^1(\Omega)$ and $f : \mathbb{R} \to \mathbb{R}$ be convex, then*

$$f\left(\frac{1}{\operatorname{meas}\Omega} \int_\Omega u(x)\, dx\right) \leq \frac{1}{\operatorname{meas}\Omega} \int_\Omega f(u(x))\, dx\,.$$

We now need to introduce the notion of duality, also known as Legendre transform, for convex functions. It is convenient to accept in the definition functions that are allowed to take the value $+\infty$ (a function that takes only finite values, is called finite).

Definition 1.53 (Legendre transform) *Let $f : \mathbb{R}^n \to \mathbb{R}$ (or $f : \mathbb{R}^n \to \mathbb{R} \cup \{+\infty\}$).*

(i) The Legendre transform, *or* dual, *of f is the function $f^* : \mathbb{R}^n \to \mathbb{R} \cup \{\pm\infty\}$ defined by*

$$f^*(x^*) = \sup_{x \in \mathbb{R}^n} \{\langle x; x^* \rangle - f(x)\}$$

where $\langle .; . \rangle$ denotes the scalar product in $\mathbb{R}^n$.

(ii) The bidual *of f is the function $f^{**} : \mathbb{R}^n \to \mathbb{R} \cup \{\pm\infty\}$ defined by*

$$f^{**}(x) = \sup_{x^* \in \mathbb{R}^n} \{\langle x; x^* \rangle - f^*(x^*)\}\,.$$

Remark 1.54 (i) In general, f^* takes the value $+\infty$, even if f takes only finite values.

(ii) If $f \not\equiv +\infty$, then $f^* > -\infty$.

Let us see some simple examples that are discussed in Exercise 1.5.4.

Example 1.55 (i) Let $n = 1$ and $f(x) = |x|^p / p$, where $1 < p < \infty$. We then find

$$f^*(x^*) = \frac{1}{p'} |x^*|^{p'}$$

where p' is, as usual, defined by $1/p + 1/p' = 1$.

(ii) Let $n = 1$ and $f(x) = (x^2 - 1)^2$. We then have

$$f^{**}(x) = \begin{cases} (x^2 - 1)^2 & \text{if } |x| \geq 1 \\ 0 & \text{if } |x| < 1. \end{cases}$$

(v) If $f : \mathbb{R}^n \to \mathbb{R}$ is strictly convex and if

$$\lim_{|x| \to \infty} \frac{f(x)}{|x|} = +\infty$$

then $f^* \in C^1(\mathbb{R}^n)$. Moreover if $f \in C^1(\mathbb{R}^n)$ and

$$f(x) + f^*(x^*) = \langle x^*; x \rangle$$

then

$$x^* = \nabla f(x) \quad and \quad x = \nabla f^*(x^*).$$

We finally conclude with a theorem that allows to compute the convex envelope without using duality (see Theorem 2.35 in [31, 2nd edition] or Corollary 17.1.5 in Rockafellar [91]).

Theorem 1.57 (Carathéodory theorem) *Let $f : \mathbb{R}^n \to \mathbb{R}$ be bounded below. Then*

$$f^{**}(x) = \inf \left\{ \sum_{i=1}^{n+1} \lambda_i f(x_i) : x = \sum_{i=1}^{n+1} \lambda_i x_i, \ \lambda_i \geq 0 \ and \ \sum_{i=1}^{n+1} \lambda_i = 1 \right\}.$$

1.5.1 Exercises

Exercise 1.5.1 Prove Theorem 1.51.

Exercise 1.5.2 Prove Jensen inequality, when $f \in C^1$.

Exercise 1.5.3 Let $f(x) = \sqrt{1 + x^2}$. Compute f^*.

Exercise 1.5.4 Establish (i), (ii) and (iv) of Example 1.55.

Exercise 1.5.5 Let $X \in \mathbb{R}^{2 \times 2}$ be a 2×2 real matrix. Show that if $f(X) = (\det X)^2$, then $f^{**}(X) \equiv 0$.

Exercise 1.5.6 Prove (i), (iii) and (iv) of Theorem 1.56. For proofs of (ii) and (v) see the bibliography in the solutions of the present exercise and the exercise below.

Exercise 1.5.7 Show (v) of Theorem 1.56 under the further restrictions that $n = 1$, $f \in C^2(\mathbb{R})$ and

$$f''(x) > 0, \ \forall x \in \mathbb{R}.$$

Prove in addition that $f^* \in C^2(\mathbb{R})$.

precise definition). However if $(u, \xi) \to f(x, u, \xi)$ is convex for every $x \in [a, b]$, then every solution of (E) is automatically a minimizer of (P).

In Section 2.3 we show that any minimizer $\overline{u}$ of (P) satisfies a different form of the Euler-Lagrange equation. Namely for every $x \in [a, b]$ the following differential equation holds:

$$\frac{d}{dx} \left[f\left(x, \overline{u}\left(x\right), \overline{u}'\left(x\right)\right) - \overline{u}'\left(x\right) f_\xi\left(x, \overline{u}\left(x\right), \overline{u}'\left(x\right)\right) \right] = f_x\left(x, \overline{u}\left(x\right), \overline{u}'\left(x\right)\right).$$

This rewriting of the equation turns out to be particularly useful when f does not depend explicitly on the variable x. Indeed we then have a first integral of (E) which is

$$f\left(\overline{u}\left(x\right), \overline{u}'\left(x\right)\right) - \overline{u}'\left(x\right) f_\xi\left(\overline{u}\left(x\right), \overline{u}'\left(x\right)\right) = \text{constant}, \ \forall x \in [a, b].$$

In Section 2.4, we present the *Hamiltonian formulation* of the problem. Roughly speaking the idea is that the solutions of (E) are also solutions (and conversely) of

$$(H) \quad \begin{cases} u'\left(x\right) = H_v\left(x, u\left(x\right), v\left(x\right)\right) \\ v'\left(x\right) = -H_u\left(x, u\left(x\right), v\left(x\right)\right) \end{cases}$$

where $v\left(x\right) = f_\xi\left(x, u\left(x\right), u'\left(x\right)\right)$ and H is the Legendre transform of f, namely

$$H\left(x, u, v\right) = \sup_{\xi \in \mathbb{R}} \left\{ v\,\xi - f\left(x, u, \xi\right) \right\}.$$

In classical mechanics f is called the *Lagrangian* and H the *Hamiltonian*.

In Section 2.5, we study the relationship between the solutions of (H) with those of a partial differential equation known as *Hamilton-Jacobi equation*

$$(HJ) \quad S_x\left(x, u\right) + H\left(x, u, S_u\left(x, u\right)\right) = 0, \ \forall\left(x, u\right) \in [a, b] \times \mathbb{R}.$$

Finally, in Section 2.6, we present the *fields theories* introduced by Weierstrass and Hilbert which allow, in certain cases, to decide if a solution of (E) is a (local or global) minimizer of (P).

We conclude this Introduction with some comments. The methods presented in this chapter can easily be generalized to vector valued functions of the form $u : [a, b] \longrightarrow \mathbb{R}^N$, with $N > 1$, to different boundary conditions, to integral constraints, or to higher derivatives. These extensions are considered in the exercises at the end of each section. However, except Section 2.2, the remaining part of the chapter does not generalize easily and completely to the multi dimensional case, $u : \Omega \subset \mathbb{R}^n \longrightarrow \mathbb{R}$, with $n > 1$; let alone the considerably harder case where $u : \Omega \subset \mathbb{R}^n \longrightarrow \mathbb{R}^N$, with $n, N > 1$.

Part 2. *Conversely if $\overline{u}$ satisfies (E) and if $(u, \xi) \to f(x, u, \xi)$ is convex for every $x \in [a, b]$ then $\overline{u}$ is a minimizer of (P).*

Part 3. *If moreover the function $(u, \xi) \to f(x, u, \xi)$ is strictly convex for every $x \in [a, b]$ then the minimizer of (P), if it exists, is unique.*

Remark 2.2 (i) One should immediately draw the attention to the fact that this theorem does not state any existence result.

(ii) As will be seen below it is not always reasonable to expect that the minimizers of (P) are $C^2([a, b])$ or even $C^1([a, b])$.

(iii) If $(u, \xi) \to f(x, u, \xi)$ is not convex (even if $\xi \to f(x, u, \xi)$ is convex for every $(x, u) \in [a, b] \times \mathbb{R}$) then a solution of (E) is not necessarily an absolute minimizer of (P). It can be a local minimizer, a local maximizer.... It is often said that such a solution of (E) is a *stationary point* of I.

(iv) The theorem easily generalizes, for example (see the exercises below), to the following cases:

- u is a vector, i.e. $u : [a, b] \to \mathbb{R}^N$, $N > 1$, the Euler-Lagrange equations are then a system of ordinary differential equations;

- $u : \Omega \subset \mathbb{R}^n \to \mathbb{R}$, $n > 1$, the Euler-Lagrange equation is then a single partial differential equation;

- $u : \Omega \subset \mathbb{R}^n \to \mathbb{R}^N$, $n, N > 1$, the Euler-Lagrange equation is then a system of partial differential equations;

- $f = f\left(x, u, u', u'', \cdots, u^{(n)}\right)$, the Euler-Lagrange equation is then an ordinary differential equation of $(2n)$th order;

- other types of boundary conditions such as $u'(a) = \alpha$, $u'(b) = \beta$;

- integral constraints of the form $\int_a^b g(x, u(x), u'(x)) \, dx = 0$.

Proof. *Part 1.* Since $\overline{u}$ is a minimizer among all elements of X, we have

$$I(\overline{u}) \leq I(\overline{u} + hv)$$

for every $h \in \mathbb{R}$ and every $v \in C^1([a, b])$ with $v(a) = v(b) = 0$. In other words, setting $\Phi(h) = I(\overline{u} + hv)$, we have that $\Phi \in C^1(\mathbb{R})$ and that $\Phi(0) \leq \Phi(h)$ for every $h \in \mathbb{R}$. We therefore deduce that

$$\Phi'(0) = \frac{d}{dh} I(\overline{u} + hv) \bigg|_{h=0} = 0$$

and hence

$$\int_a^b \left[f_\xi(x, \overline{u}(x), \overline{u}'(x)) \, v'(x) + f_u(x, \overline{u}(x), \overline{u}'(x)) \, v(x) \right] dx = 0. \tag{2.1}$$

We therefore get

$$\int_a^b \left[\frac{1}{2} f\left(x, u, u'\right) + \frac{1}{2} f\left(x, v, v'\right) - f\left(x, \frac{1}{2}u + \frac{1}{2}v, \frac{1}{2}u' + \frac{1}{2}v'\right) \right] dx = 0 \,.$$

Since the integrand is, by strict convexity of f, positive unless $u = v$ and $u' = v'$ we deduce that $u \equiv v$, as wished. ∎

We now consider several particular cases and examples that are arranged in an order of increasing difficulty.

Case 2.3 $f\left(x, u, \xi\right) = f\left(\xi\right)$.

This is the simplest case. The Euler-Lagrange equation is

$$\frac{d}{dx}\left[f'\left(u'\right)\right] = 0, \ \text{i.e.} \ f'\left(u'\right) = \text{constant}.$$

Note that

$$\bar{u}\left(x\right) = \frac{\beta - \alpha}{b - a}\left(x - a\right) + \alpha \tag{2.2}$$

is a solution of the equation and furthermore satisfies the boundary conditions $\bar{u}\left(a\right) = \alpha$, $\bar{u}\left(b\right) = \beta$. It is therefore a stationary point of I. It is not, however, always a minimizer of (P) as will be seen in the second and third examples.

1. **f is convex.**

    If f is convex, the above $\bar{u}$ is indeed a minimizer. This follows from the theorem but it can be seen in a more elementary way (which is also valid even if $f \in C^0\left(\mathbb{R}\right)$). From Jensen inequality (cf. Theorem 1.52) it follows that for any $u \in C^1\left([a, b]\right)$ with $u\left(a\right) = \alpha$, $u\left(b\right) = \beta$

$$\frac{1}{b - a} \int_a^b f\left(u'\left(x\right)\right) dx \geq f\left(\frac{1}{b - a} \int_a^b u'\left(x\right) dx\right) = f\left(\frac{u\left(b\right) - u\left(a\right)}{b - a}\right)$$

$$= f\left(\frac{\beta - \alpha}{b - a}\right) = f\left(\bar{u}'\left(x\right)\right)$$

$$= \frac{1}{b - a} \int_a^b f\left(\bar{u}'\left(x\right)\right) dx$$

    which is the claim. If f is not strictly convex, then, in general, there are other minimizers (see Exercise 2.2.9).

2. **f is non-convex.**

This last problem has clearly

$$v\left(x\right) = \begin{cases} x & \text{if } x \in [0, 1/2] \\ 1 - x & \text{if } x \in (1/2, 1] \end{cases}$$

as a solution since v is *piecewise* C^1 and satisfies $v\left(0\right) = v\left(1\right) = 0$ and $I\left(v\right) = 0$; thus $m_{\text{piec}} = 0$. Assume for a moment that we already proved that not only $m_{\text{piec}} = 0$ but also $m = 0$. This readily implies that (P), contrary to (P_{piec}), has no solution. Indeed $I\left(u\right) = 0$ implies that $|u'| = 1$ almost everywhere and no function $u \in X$ can satisfy $|u'| = 1$ (since by continuity of the derivative we should have either $u' = 1$ everywhere or $u' = -1$ everywhere and this is incompatible with the boundary data).

We now prove that $m = 0$. We give a direct argument now and a more elaborate one in Exercise 2.2.6. Consider the following sequence

$$u_\nu\left(x\right) = \begin{cases} x & \text{if } x \in \left[0, \frac{1}{2} - \frac{1}{\nu}\right] \\ -2\nu^2 \left(x - \frac{1}{2}\right)^3 - 4\nu \left(x - \frac{1}{2}\right)^2 - x + 1 & \text{if } x \in \left(\frac{1}{2} - \frac{1}{\nu}, \frac{1}{2}\right] \\ 1 - x & \text{if } x \in \left(\frac{1}{2}, 1\right]. \end{cases}$$

Observe that $u_\nu \in X$ and

$$I\left(u_\nu\right) = \int_0^1 f\left(u_\nu'\left(x\right)\right) dx = \int_{\frac{1}{2} - \frac{1}{\nu}}^{\frac{1}{2}} f\left(u_\nu'\left(x\right)\right) dx \leq \frac{4}{\nu} \to 0.$$

This implies that indeed $m = 0$.

We can also make the further observation that the Euler-Lagrange equation is

$$\frac{d}{dx} \left[u' \left(\left(u'\right)^2 - 1 \right) \right] = 0.$$

It has $\overline{u} \equiv 0$ as a solution. However, since $m = 0$, it is not a minimizer $(I\left(0\right) = 1)$.

Case 2.4 $f\left(x, u, \xi\right) = f\left(x, \xi\right).$

The Euler-Lagrange equation is

$$\frac{d}{dx} \left[f_\xi\left(x, u'\right) \right] = 0, \text{ i.e. } f_\xi\left(x, u'\right) = \text{constant}.$$

The equation is already harder to solve than the preceding one and, in general, it has not a solution as simple as the one in (2.2).

We finally prove that $m = 0$. This can be done in two different ways. A more sophisticated argument is given in Exercise 2.2.6 and it provides an interesting continuity argument. A possible approach is to consider the following sequence

$$u_\nu(x) = \begin{cases} \frac{-\nu^2}{\log \nu} x^2 + \frac{\nu}{\log \nu} x + 1 & \text{if } x \in \left[0, \frac{1}{\nu}\right] \\ \frac{-\log x}{\log \nu} & \text{if } x \in \left(\frac{1}{\nu}, 1\right]. \end{cases}$$

We easily have $u_\nu \in X$ and since

$$u_\nu'(x) = \begin{cases} \frac{\nu}{\log \nu}(1 - 2\nu x) & \text{if } x \in \left[0, \frac{1}{\nu}\right] \\ \frac{-1}{x \log \nu} & \text{if } x \in \left(\frac{1}{\nu}, 1\right] \end{cases}$$

we deduce that

$$0 \leq I(u_\nu) = \frac{\nu^2}{\log^2 \nu} \int_0^{1/\nu} x(1 - 2\nu x)^2 \, dx + \frac{1}{\log^2 \nu} \int_{1/\nu}^1 \frac{dx}{x} \to 0, \text{ as } \nu \to \infty.$$

This indeed shows that $m = 0$.

Case 2.5 $f(x, u, \xi) = f(u, \xi)$.

Although this case is a lot harder to treat than the preceding ones it has an important property that is not present in the most general case when $f = f(x, u, \xi)$. The Euler-Lagrange equation is

$$\frac{d}{dx}[f_\xi(u(x), u'(x))] = f_u(u(x), u'(x)), \quad x \in (a, b)$$

and according to Theorem 2.8 below, it has a *first integral* that is given by

$$f(u(x), u'(x)) - u'(x) f_\xi(u(x), u'(x)) = \text{constant}, \quad x \in (a, b).$$

1. **Poincaré-Wirtinger inequality**.

We will show, in several steps, that

$$\int_a^b u'^2 \, dx \geq \left(\frac{\pi}{b - a}\right)^2 \int_a^b u^2 \, dx$$

for every u satisfying $u(a) = u(b) = 0$. By a change of variable we immediately reduce the study to the case $a = 0$ and $b = 1$. We will also prove in Theorem 6.1 a slightly more general inequality known as *Wirtinger inequality* which states that

$$\int_{-1}^1 u'^2 \, dx \geq \pi^2 \int_{-1}^1 u^2 \, dx$$

The Euler-Lagrange equation and its first integral are

$$\left[\frac{u'}{\sqrt{u}\sqrt{1+u'^2}}\right]' = -\frac{\sqrt{1+u'^2}}{2\sqrt{u^3}}$$

$$\frac{\sqrt{1+u'^2}}{\sqrt{u}} - u'\left[\frac{u'}{\sqrt{u}\sqrt{1+u'^2}}\right] = \text{constant.}$$

This leads (μ being a positive constant) to

$$u\left(1+u'^2\right) = 2\mu.$$

The solution is a *cycloid* and it is given in implicit form by

$$u(x) = \mu\left(1 - \cos\theta^{-1}(x)\right)$$

where

$$\theta(t) = \mu(t - \sin t).$$

Note that $u(0) = 0$. It therefore remains to choose μ so that $u(b) = \beta$.

3. **Minimal surfaces of revolution.**

This example is treated in Chapter 5. Let us briefly present it here. The function under consideration is $f(u,\xi) = 2\pi u\sqrt{1+\xi^2}$ and the minimization problem (which corresponds to minimization of the area of a surface of revolution) is

$$(P) \quad \inf_{u \in X}\left\{I(u) = \int_a^b f(u(x), u'(x))\,dx\right\} = m$$

where

$$X = \left\{u \in C^1([a,b]) : u(a) = \alpha,\ u(b) = \beta,\ u > 0\right\}$$

and $\alpha,\ \beta > 0$. The Euler-Lagrange equation and its first integral are

$$\left[\frac{u'u}{\sqrt{1+u'^2}}\right]' = \sqrt{1+u'^2} \Leftrightarrow u''u = 1+u'^2$$

$$u\sqrt{1+u'^2} - u'\frac{u'u}{\sqrt{1+u'^2}} = \lambda = \text{constant.}$$

This leads to

$$u'^2 = \frac{u^2}{\lambda^2} - 1.$$

The solutions, if they exist (this depends on a, b, α and β, see Exercise 5.2.3), are of the form (μ being a constant)

$$u(x) = \lambda\cosh\left(\frac{x}{\lambda} + \mu\right).$$

2.2.1   Exercises

Exercise 2.2.1 Generalize Theorem 2.1 to the case where $u : [a, b] \to \mathbb{R}^N$, $N \geq 1$.

Exercise 2.2.2 Generalize Theorem 2.1 to the case where $u : [a, b] \to \mathbb{R}$ and

$$(P) \quad \inf_{u \in X} \left\{ I(u) = \int_a^b f\left(x, u(x), \cdots, u^{(n)}(x)\right) dx \right\}$$

where $X = \left\{ u \in C^n([a, b]) : u^{(j)}(a) = \alpha_j, \ u^{(j)}(b) = \beta_j, \ 0 \leq j \leq n - 1 \right\}.$

Exercise 2.2.3 **(i)** Find the appropriate formulation of Theorem 2.1 when $u : [a, b] \to \mathbb{R}$ and

$$(P) \quad \inf_{u \in X} \left\{ I(u) = \int_a^b f(x, u(x) u'(x)) dx \right\}$$

where $X = \left\{ u \in C^1([a, b]) : u(a) = \alpha \right\}$, i.e. we leave one of the end points free.

(ii) Similar question, when we leave both end points free; i.e. when we minimize I over $C^1([a, b])$.

Exercise 2.2.4 (Lagrange multiplier) Generalize Theorem 2.1 in the following case where $u : [a, b] \to \mathbb{R}$,

$$(P) \quad \inf_{u \in X} \left\{ I(u) = \int_a^b f(x, u(x), u'(x)) dx \right\},$$

$$X = \left\{ u \in C^1([a, b]) : u(a) = \alpha, \ u(b) = \beta, \ \int_a^b g(x, u(x), u'(x)) dx = 0 \right\}$$

where $g \in C^2([a, b] \times \mathbb{R} \times \mathbb{R})$.

Exercise 2.2.5 (Second variation of I) Let $f \in C^3([a, b] \times \mathbb{R} \times \mathbb{R})$ and

$$(P) \quad \inf_{u \in X} \left\{ I(u) = \int_a^b f(x, u(x), u'(x)) dx \right\}$$

where $X = \left\{ u \in C^1([a, b]) : u(a) = \alpha, \ u(b) = \beta \right\}$. Let $\overline{u} \in X \cap C^2([a, b])$ be a minimizer for (P). Show that the following inequality

$$\int_a^b \left[f_{uu}(x, \overline{u}, \overline{u}') v^2 + 2 f_{u\xi}(x, \overline{u}, \overline{u}') vv' + f_{\xi\xi}(x, \overline{u}, \overline{u}') v'^2 \right] dx \geq 0$$

Exercise 2.2.9 Let $X = \{u \in C^1([0,1]) : u(0) = 0, \ u(1) = 1\}$ and

$$(P) \quad \inf_{u \in X} \left\{ I(u) = \int_0^1 |u'(x)| \, dx \right\} = m.$$

Prove that (P) has infinitely many solutions.

Exercise 2.2.10 Let $p \geq 1$ and $a \in C^0(\mathbb{R})$, with $a(u) \geq a_0 > 0$. Let A be defined by

$$A'(u) = [a(u)]^{1/p}.$$

Show that a minimizer (which is unique if $p > 1$) of

$$(P) \quad \inf_{u \in X} \left\{ I(u) = \int_a^b a(u(x)) |u'(x)|^p \, dx \right\}$$

where $X = \{u \in C^1([a,b]) : u(a) = \alpha, \ u(b) = \beta\}$ is given by

$$u(x) = A^{-1} \left[\frac{A(\beta) - A(\alpha)}{b - a} (x - a) + A(\alpha) \right].$$

2.3 Second form of the Euler-Lagrange equation

The next theorem gives a different way of expressing the Euler-Lagrange equation, this new equation is sometimes called *DuBois-Reymond equation*. It turns out to be useful when f does not depend explicitly on x, as already seen in some of the above examples.

Theorem 2.8 *Let* $f \in C^2([a,b] \times \mathbb{R} \times \mathbb{R})$, $f = f(x, u, \xi)$, *and*

$$(P) \quad \inf_{u \in X} \left\{ I(u) = \int_a^b f(x, u(x), u'(x)) \, dx \right\} = m$$

where $X = \{u \in C^1([a,b]) : u(a) = \alpha, \ u(b) = \beta\}$. *Let* $u \in X \cap C^2([a,b])$ *be a minimizer of* (P) *then for every* $x \in [a,b]$ *the following equation holds*

$$\frac{d}{dx} [f(x, u(x), u'(x)) - u'(x) f_\xi(x, u(x), u'(x))] = f_x(x, u(x), u'(x)). \quad (2.3)$$

Proof. We give two different proofs of the theorem. The first one is very elementary and uses the Euler-Lagrange equation. The second one is more involved but has several advantages that we do not discuss now.

Note that, performing also a change of variables $y = \xi(x, \epsilon)$,

$$
\begin{aligned}
I(u^\epsilon) &= \int_a^b f\left(x, u^\epsilon(x), (u^\epsilon)'(x)\right) dx \\
&= \int_a^b f\left(x, u(\xi(x, \epsilon)), u'(\xi(x, \epsilon)) \xi_x(x, \epsilon)\right) dx \\
&= \int_a^b f\left(\eta(y, \epsilon), u(y), u'(y)/\eta_y(y, \epsilon)\right) \eta_y(y, \epsilon) \, dy.
\end{aligned}
$$

Denoting by $g(\epsilon)$ the last integrand, we get

$$
g'(\epsilon) = \eta_{y\epsilon} f + \left[f_x \eta_\epsilon - \frac{\eta_{y\epsilon}}{\eta_y^2} u' f_\xi\right] \eta_y
$$

which leads to

$$
g'(0) = \lambda\left[-f_x \varphi + (u' f_\xi - f)\varphi'\right].
$$

Since by hypothesis u is a minimizer of (P) and $u^\epsilon \in X$ we have $I(u^\epsilon) \geq I(u)$ and hence

$$
\begin{aligned}
0 = \frac{d}{d\epsilon} I(u^\epsilon)\bigg|_{\epsilon=0} &= \lambda \int_a^b \{-f_x\left(x, u(x), u'(x)\right)\varphi(x) \\
&+ \left[u'(x) f_\xi\left(x, u(x), u'(x)\right) - f\left(x, u(x), u'(x)\right)\right]\varphi'(x)\} \, dx \\
&= \lambda \int_a^b \{-f_x\left(x, u(x), u'(x)\right) \\
&+ \frac{d}{dx}\left[-u'(x) f_\xi\left(x, u(x), u'(x)\right) + f\left(x, u(x), u'(x)\right)\right]\}\varphi(x) \, dx.
\end{aligned}
$$

Appealing, once more, to Theorem 1.24 we have indeed obtained the claim. ∎

2.3.1   Exercises

Exercise 2.3.1 Generalize Theorem 2.8 to the case where $u : [a, b] \to \mathbb{R}^N$, $N \geq 1$.

Exercise 2.3.2 Let

$$
f(x, u, \xi) = f(u, \xi) = \frac{1}{2}\xi^2 - u.
$$

Show that $u \equiv 1$ is a solution of (2.3), but not of the Euler-Lagrange equation (E).

Then $H \in C^2 ([a, b] \times \mathbb{R} \times \mathbb{R})$ and

$$H_x (x, u, v) = -f_x (x, u, H_v (x, u, v)) \tag{2.7}$$

$$H_u (x, u, v) = -f_u (x, u, H_v (x, u, v)) \tag{2.8}$$

$$H (x, u, v) = v H_v (x, u, v) - f (x, u, H_v (x, u, v)) \tag{2.9}$$

$$v = f_\xi (x, u, \xi) \Leftrightarrow \xi = H_v (x, u, v). \tag{2.10}$$

Remark 2.10 (i) The lemma remains partially true if we replace the hypothesis (2.4) by the weaker condition

$$\xi \rightarrow f (x, u, \xi) \text{ is strictly convex.}$$

In general, however the function H is only C^1, as the following simple example shows

$$f (x, u, \xi) = \frac{1}{4} |\xi|^4 \quad \text{and} \quad H (x, u, v) = \frac{3}{4} |v|^{4/3}.$$

(See also Example 2.14.)

(ii) The lemma also remains valid if the hypothesis (2.5) does not hold but then, in general, H is no longer finite everywhere as the following simple example suggests. Consider the strictly convex function

$$f (x, u, \xi) = f (\xi) = \sqrt{1 + \xi^2}$$

and observe (cf. Exercise 1.5.3) that

$$H (v) = \begin{cases} -\sqrt{1 - v^2} & \text{if } |v| \leq 1 \\ +\infty & \text{if } |v| > 1. \end{cases}$$

(iii) The same proof leads to the fact that if $f \in C^k$, $k \geq 2$, then $H \in C^k$.

Proof. We divide the proof into four steps.

Step 1. Fix $(x, u, v) \in [a, b] \times \mathbb{R} \times \mathbb{R}$. We first show that the supremum, in the definition of H, is attained. Assume, for the sake of contradiction, that there exists a sequence $\xi_\nu \in \mathbb{R}$ with $|\xi_\nu| \rightarrow \infty$ such that

$$-f (x, u, 0) \leq H (x, u, v) \leq \frac{1}{\nu} + v\xi_\nu - f (x, u, \xi_\nu).$$

Use (2.5) to get

$$-f (x, u, 0) \leq H (x, u, v) \leq \frac{1}{\nu} + |\xi_\nu| \left[v \frac{\xi_\nu}{|\xi_\nu|} - \frac{\omega (|\xi_\nu|)}{|\xi_\nu|} \right] - g (x, u).$$

Step 3. The inverse function theorem, the fact that $f \in C^2$ and the inequality (2.4) imply that $\xi \in C^1$. But, as an exercise, we establish this fact again. First let us prove that ξ is continuous (in fact locally Lipschitz). Let $R > 0$ be fixed and $R_1 > 0$ so that (cf. Step 1)

$$|\xi(x,u,v)| \leq R_1, \text{ for every } x \in [a,b], \ |u|, |v| \leq R.$$

Since f_ξ is C^1, we can find $\gamma_1 > 0$ so that

$$|f_\xi(x,u,\xi) - f_\xi(x',u',\xi')| \leq \gamma_1 (|x - x'| + |u - u'| + |\xi - \xi'|) \qquad (2.13)$$

for every x, $x' \in [a,b]$, $|u|$, $|u'| \leq R$, $|\xi|$, $|\xi'| \leq R_1$.

From (2.4), we find that there exists $\gamma_2 > 0$ so that

$$f_{\xi\xi}(x,u,\xi) \geq \gamma_2, \text{ for every } x \in [a,b], \ |u| \leq R, \ |\xi| \leq R_1$$

and we thus have, for every $x \in [a,b]$, $|u| \leq R$, $|\xi|$, $|\xi'| \leq R_1$,

$$|f_\xi(x,u,\xi) - f_\xi(x,u,\xi')| \geq \gamma_2 |\xi - \xi'|. \qquad (2.14)$$

Let x, $x' \in [a,b]$, $|u|$, $|u'| \leq R$, $|v|$, $|v'| \leq R$. By definition of ξ we have

$$f_\xi(x,u,\xi(x,u,v)) = v$$

$$f_\xi(x',u',\xi(x',u',v')) = v',$$

which leads to

$$f_\xi(x,u,\xi(x',u',v')) - f_\xi(x,u,\xi(x,u,v))$$
$$= f_\xi(x,u,\xi(x',u',v')) - f_\xi(x',u',\xi(x',u',v')) + v' - v.$$

Combining this identity with (2.13) and (2.14) we get

$$\gamma_2 |\xi(x,u,v) - \xi(x',u',v')| \leq \gamma_1 (|x - x'| + |u - u'|) + |v - v'|$$

which, indeed, establishes the continuity of ξ.

We now show that ξ is in fact C^1. From the equation $v = f_\xi(x,u,\xi)$ we deduce that

$$\begin{cases} f_{x\xi}(x,u,\xi) + f_{\xi\xi}(x,u,\xi)\,\xi_x = 0 \\ f_{u\xi}(x,u,\xi) + f_{\xi\xi}(x,u,\xi)\,\xi_u = 0 \\ \qquad\quad f_{\xi\xi}(x,u,\xi)\,\xi_v = 1. \end{cases}$$

Since (2.4) holds and $f \in C^2$, we deduce that $\xi \in C^1([a,b] \times \mathbb{R} \times \mathbb{R})$.

Step 4. We therefore have that the functions

$$(x,u,v) \to \xi(x,u,v), \ f_x(x,u,\xi(x,u,v)), \ f_u(x,u,\xi(x,u,v))$$

Example 2.13 The present example is motivated by classical mechanics. Let $m > 0$, $g \in C^1([a, b])$ and

$$f(x, u, \xi) = \frac{m}{2}\xi^2 - g(x)u.$$

The integral under consideration is

$$I(u) = \int_a^b f(x, u(x), u'(x))\, dx$$

and the associated Euler-Lagrange equation is

$$mu''(x) = -g(x), \quad x \in (a, b).$$

The Hamiltonian is then

$$H(x, u, v) = \frac{v^2}{2m} + g(x)u$$

while the associated Hamiltonian system is

$$\begin{cases} u'(x) = v(x)/m \\ v'(x) = -g(x). \end{cases}$$

Example 2.14 We now generalize the preceding example. Let $p > 1$ and $p' = p/(p-1)$,

$$f(x, u, \xi) = \frac{1}{p}|\xi|^p - g(x, u) \quad \text{and} \quad H(x, u, v) = \frac{1}{p'}|v|^{p'} + g(x, u).$$

The Euler-Lagrange equation and the associated Hamiltonian system are

$$\frac{d}{dx}\left[|u'|^{p-2}u'\right] = -g_u(x, u)$$

and

$$\begin{cases} u' = |v|^{p'-2}v \\ v' = -g_u(x, u). \end{cases}$$

Example 2.15 Consider the simplest case where $f(x, u, \xi) = f(\xi)$ with $f'' > 0$ (or more generally f is strictly convex) and $\lim_{|\xi| \to \infty} f(\xi)/|\xi| = +\infty$. The Euler-Lagrange equation and its integrated form are

$$\frac{d}{dx}[f'(u')] = 0 \Rightarrow f'(u') = \lambda = \text{constant}.$$

The associated Hamiltonian system is

$$\begin{cases} u'(x) = H_v(u(x), v(x)) \\ v'(x) = -H_u(u(x), v(x)). \end{cases}$$

The Hamiltonian system also has a first integral given by

$$\frac{d}{dx}[H(u,v)] = H_u(u,v)u' + H_v(u,v)v' \equiv 0.$$

In physical terms we can say that if the Lagrangian f is independent of the variable x (which is here the time), the Hamiltonian H is constant along the trajectories.

2.4.1   Exercises

Exercise 2.4.1 Generalize Theorem 2.11 to the case where $u : [a,b] \to \mathbb{R}^N$, $N \geq 1$.

Exercise 2.4.2 Consider a mechanical system with N particles whose respective masses are m_i and positions at time t are $u_i(t) = (x_i(t), y_i(t), z_i(t)) \in \mathbb{R}^3$, $1 \leq i \leq N$. Let

$$T(u') = \frac{1}{2}\sum_{i=1}^N m_i |u_i'|^2 = \frac{1}{2}\sum_{i=1}^N m_i \left(x_i'^2 + y_i'^2 + z_i'^2\right)$$

be the kinetic energy and denote by $U = U(t,u)$ the potential energy. Finally let

$$L(t,u,u') = T(u') - U(t,u)$$

be the Lagrangian. Let also H be the associated Hamiltonian. With the help of the preceding exercise show the following results.

(i) Write the Euler-Lagrange equations. Find the associated Hamiltonian system.

(ii) Show that, along the trajectories (i.e. when $v = L_\xi(t,u,u')$), the Hamiltonian can be written as (in mechanical terms it is the total energy of the system)

$$H(t,u,v) = T(u') + U(t,u).$$

Exercise 2.4.3 Let $f(x,u,\xi) = \sqrt{g(x,u)}\sqrt{1+\xi^2}$. Write the associated Hamiltonian system and find a first integral of this system when g does not depend explicitly on x.

leads immediately to a solution of (2.15), setting

$$S(x, u, \alpha) = S^*(u, \alpha) - \alpha x.$$

(ii) It is, in general, a difficult task to solve (2.15) and an extensive bibliography on the subject exists, cf. Evans [44], Lions [73].

Proof. *Step 1.* We differentiate (2.17) to get

$$v'(x) = S_{xu}(x, u(x)) + u'(x) S_{uu}(x, u(x)), \ \forall x \in [a, b].$$

Differentiating (2.15) with respect to u we find, for every $(x, u) \in [a, b] \times \mathbb{R}$,

$$S_{xu}(x, u) + H_u(x, u, S_u(x, u)) + H_v(x, u, S_u(x, u)) S_{uu}(x, u) = 0.$$

Combining the two identities (the second one evaluated at $u = u(x)$) and (2.16) with the definition of v, we have

$$v'(x) = -H_u(x, u(x), S_u(x, u(x))) = -H_u(x, u(x), v(x))$$

as wished.

Step 2. Since S is a solution of the Hamilton-Jacobi equation, we have, for every $(x, u, \alpha) \in [a, b] \times \mathbb{R} \times \mathbb{R}$,

$$\frac{d}{d\alpha} [S_x(x, u, \alpha) + H(x, u, S_u(x, u, \alpha))]$$
$$= S_{x\alpha}(x, u, \alpha) + H_v(x, u, S_u(x, u, \alpha)) S_{u\alpha}(x, u, \alpha) = 0.$$

Since this identity is valid for every u, it is also valid for $u = u(x)$ satisfying (2.16) and thus

$$S_{x\alpha}(x, u(x), \alpha) + u'(x) S_{u\alpha}(x, u(x), \alpha) = 0.$$

This last identity can be rewritten as

$$\frac{d}{dx} [S_\alpha(x, u(x), \alpha)] = 0$$

which is the claim. ∎

The above theorem admits a converse.

Theorem 2.20 (Jacobi theorem) *Let* $H \in C^1([a, b] \times \mathbb{R} \times \mathbb{R})$, $S = S(x, u, \alpha)$ *be* $C^2([a, b] \times \mathbb{R} \times \mathbb{R})$ *with*

$$S_{u\alpha}(x, u, \alpha) \neq 0, \ \forall (x, u, \alpha) \in [a, b] \times \mathbb{R} \times \mathbb{R}$$

Combining the two identities (the second one evaluated at $u = u(x)$) we infer the result, namely

$$v'(x) = -H_u(x, u(x), S_u(x, u(x), \alpha)) = -H_u(x, u(x), v(x)).$$

This achieves the proof of the theorem. ∎

Example 2.21 Let $g \in C^1(\mathbb{R})$ with $g(u) \geq g_0 > 0$. Let

$$H(u, v) = \frac{1}{2}v^2 - g(u)$$

be the Hamiltonian associated to

$$f(u, \xi) = \frac{1}{2}\xi^2 + g(u).$$

The Hamilton-Jacobi equation and its reduced form (cf. (2.19)) are given by

$$S_x + \frac{1}{2}(S_u)^2 - g(u) = 0 \quad \text{and} \quad \frac{1}{2}(S_u)^2 = g(u).$$

Therefore a solution of the equation is given by

$$S = S(x, u) = S(u) = \int_0^u \sqrt{2g(s)}\, ds.$$

We next solve

$$u'(x) = H_v(u(x), S_u(u(x))) = S_u(u(x)) = \sqrt{2g(u(x))}$$

which has a solution given implicitly by

$$\int_{u(0)}^{u(x)} \frac{ds}{\sqrt{2g(s)}} = x.$$

Setting $v(x) = S_u(u(x))$, we have indeed found a solution of the Hamiltonian system

$$\begin{cases} u'(x) = H_v(u(x), v(x)) = v(x) \\ v'(x) = -H_u(u(x), v(x)) = g'(u(x)). \end{cases}$$

Note also that such a function u solves

$$u''(x) = g'(u(x))$$

which is the Euler-Lagrange equation associated to the Lagrangian f.

where

$$\widetilde{f}(x, u, \xi) = f(x, u, \xi) + \Phi_u(x, u)\xi + \Phi_x(x, u);$$

then any solution $\overline{u}$ *of* (E) *is a minimizer of* (P).

Remark 2.23 We should immediately point out that in order to have $(u, \xi) \to \widetilde{f}(x, u, \xi)$ convex for every $x \in [a, b]$ we should, at least, have that $\xi \to f(x, u, \xi)$ is convex for every $(x, u) \in [a, b] \times \mathbb{R}$. If $(u, \xi) \to f(x, u, \xi)$ is already convex, then choose $\Phi \equiv 0$ and apply Theorem 2.1.

Proof. Define

$$\varphi(x, u, \xi) = \Phi_u(x, u)\xi + \Phi_x(x, u).$$

Observe that the following two identities (the first one uses that $\Phi(a, \alpha) = \Phi(b, \beta)$ and the second one is just straight differentiation)

$$\int_a^b \frac{d}{dx}[\Phi(x, u(x))]\, dx = \Phi(b, \beta) - \Phi(a, \alpha) = 0$$

$$\frac{d}{dx}[\varphi_\xi(x, u, u')] = \varphi_u(x, u, u'), \quad x \in [a, b]$$

hold for any $u \in X = \{u \in C^1([a, b]) : u(a) = \alpha,\ u(b) = \beta\}$. The first identity expresses that the integral is *invariant*, while the second one says that $\varphi(x, u, u')$ satisfies the Euler-Lagrange equation identically (it is then called a *null Lagrangian*).

With the help of the above observations we immediately obtain the result by applying Theorem 2.1 to $\widetilde{f}$. Indeed we have that $(u, \xi) \to \widetilde{f}(x, u, \xi)$ is convex,

$$I(u) = \int_a^b \widetilde{f}(x, u(x), u'(x))\, dx = \int_a^b f(x, u(x), u'(x))\, dx$$

for every $u \in X$ and any solution $\overline{u}$ of (E) also satisfies

$$\left(\widetilde{E}\right) \quad \frac{d}{dx}\left[\widetilde{f}_\xi(x, \overline{u}, \overline{u}')\right] = \widetilde{f}_u(x, \overline{u}, \overline{u}'), \quad x \in (a, b).$$

This concludes the proof. ∎

With the help of the above elementary theorem we can now fully handle the Poincaré-Wirtinger inequality.

Example 2.24 (Poincaré-Wirtinger inequality) Let $\lambda \geq 0$,

$$f_\lambda(u, \xi) = \frac{\xi^2 - \lambda^2 u^2}{2}$$

We start with an elementary result that is a first justification for defining such a notion.

Proposition 2.27 *Let* $f \in C^2 \left([a, b] \times \mathbb{R} \times \mathbb{R}\right)$, $f = f\left(x, u, \xi\right)$, *and*

$$I\left(u\right) = \int_a^b f\left(x, u\left(x\right), u'\left(x\right)\right) dx.$$

Let $\Phi : D \to \mathbb{R}^2$, $\Phi = \Phi\left(x, u\right)$ *be an exact field for* f *covering* D, $[a, b] \times \mathbb{R} \subset D$. *Then any solution* $u \in C^2 \left([a, b]\right)$ *of*

$$u'\left(x\right) = \Phi\left(x, u\left(x\right)\right) \tag{2.21}$$

solves the Euler-Lagrange associated to the functional I, *namely*

$$(E) \quad \frac{d}{dx} \left[f_\xi\left(x, u\left(x\right), u'\left(x\right)\right)\right] = f_u\left(x, u\left(x\right), u'\left(x\right)\right), \ x \in [a, b]. \tag{2.22}$$

Proof. By definition of Φ and using the fact that $p = f_\xi$, we have, for any $(x, u) \in D$,

$$h_u = f_u\left(x, u, \Phi\right) + f_\xi\left(x, u, \Phi\right) \Phi_u - p_u \Phi - p \Phi_u = f_u\left(x, u, \Phi\right) - p_u \Phi$$

and hence

$$f_u\left(x, u, \Phi\right) = h_u\left(x, u\right) + p_u\left(x, u\right) \Phi\left(x, u\right).$$

We therefore get for every $x \in [a, b]$

$$\frac{d}{dx} \left[f_\xi\left(x, u, u'\right)\right] - f_u\left(x, u, u'\right) = \frac{d}{dx} \left[p\left(x, u\right)\right] - \left[h_u\left(x, u\right) + p_u\left(x, u\right) \Phi\left(x, u\right)\right]$$

$$= p_x + p_u u' - h_u - p_u \Phi = p_x - h_u = 0$$

since we have that $u' = \Phi$ and $p_x = h_u$, Φ being exact. Thus we have reached the claim. ∎

The next theorem is the main result of this section and was established by Weierstrass and Hilbert.

Theorem 2.28 (Hilbert theorem) *Let* $f \in C^2 \left([a, b] \times \mathbb{R} \times \mathbb{R}\right)$ *with*

$$\xi \to f\left(x, u, \xi\right) \ convex \ for \ every \ (x, u) \in [a, b] \times \mathbb{R}.$$

Let $D \subset \mathbb{R}^2$ *be a domain and* $\Phi : D \to \mathbb{R}^2$, $\Phi = \Phi\left(x, u\right)$, *be an exact field for* f *covering* D. *Assume that there exists* $u_0 \in C^1 \left([a, b]\right)$ *satisfying*

$$\left(x, u_0\left(x\right)\right) \in D \quad and \quad u_0'\left(x\right) = \Phi\left(x, u_0\left(x\right)\right), \ \forall x \in [a, b],$$

Since $E\left(x, u_0, \Phi\left(x, u_0\right), u_0'\right) = 0$ we have that

$$I\left(u_0\right) = S\left(b, u_0\left(b\right)\right) - S\left(a, u_0\left(a\right)\right).$$

Moreover since $u_0\left(a\right) = u\left(a\right)$, $u_0\left(b\right) = u\left(b\right)$ we deduce that $I\left(u\right) \geq I\left(u_0\right)$ for every $u \in X$. This achieves the proof of the theorem.

The quantity

$$\int_a^b \frac{d}{dx}\left[S\left(x, u\left(x\right)\right)\right] dx$$

is called *invariant Hilbert integral.* ∎

2.6.1   Exercises

Exercise 2.6.1 Generalize Theorem 2.22 to the case where $u : [a, b] \to \mathbb{R}^N$, $N \geq 1$.

Exercise 2.6.2 Generalize Hilbert Theorem (Theorem 2.28) to the case where $u : [a, b] \to \mathbb{R}^N$, $N \geq 1$.

Exercise 2.6.3 (The present exercise establishes the connection between exact field and Hamilton-Jacobi equation.) Let $f = f\left(x, u, \xi\right)$ and $H = H\left(x, u, v\right)$ be as in Theorem 2.11 and Lemma 2.9.

(i) Show that if there exists an exact field Φ covering D, then

$$S_x + H\left(x, u, S_u\right) = 0, \ \forall\left(x, u\right) \in D$$

where

$$S_u\left(x, u\right) = f_\xi\left(x, u, \Phi\left(x, u\right)\right)$$

$$S_x\left(x, u\right) = f\left(x, u, \Phi\left(x, u\right)\right) - S_u\left(x, u\right)\Phi\left(x, u\right).$$

(ii) Conversely if the Hamilton-Jacobi equation has a solution for every $\left(x, u\right) \in D$, prove that

$$\Phi\left(x, u\right) = H_v\left(x, u, S_u\left(x, u\right)\right)$$

is an exact field for f covering D.

satisfies (H_1) but verifies (H_2) only with $p = 1$. This problem requires a special treatment (see Chapter 5).

It is interesting to compare the generality of the result with those of the preceding chapter. The main drawback of the present analysis is that we prove existence of minima only in Sobolev spaces. In the next chapter we will see that, under some extra hypotheses, the solution is in fact more regular (for example it is C^1, C^2 or C^∞).

We now describe the content of the present chapter. In Section 3.2 we consider the model case, namely the Dirichlet integral. Although this is just an example of the more general theorem obtained in Section 3.3, we fully discuss the particular case because of its importance and to make easier the understanding of the method. Recall that the origin of the direct methods goes back to Hilbert, Lebesgue and Tonelli, while treating the Dirichlet integral. Let us briefly describe the two main steps in the proof.

Step 1 (Compactness). Let $u_\nu \in u_0 + W_0^{1,p}(\Omega)$ be a minimizing sequence of (P), this means that

$$I(u_\nu) \to \inf \{I(u)\} = m, \text{ as } \nu \to \infty.$$

It is easy, invoking (H_2) and Poincaré inequality (cf. Theorem 1.48), to obtain that there exist $\overline{u} \in u_0 + W_0^{1,p}(\Omega)$ and a subsequence (still denoted u_ν) so that u_ν converges weakly to $\overline{u}$ in $W^{1,p}$, i.e.

$$u_\nu \rightharpoonup \overline{u} \text{ in } W^{1,p}, \text{ as } \nu \to \infty.$$

Step 2 (Lower semicontinuity). We then show that (H_1) implies the (sequential) weak lower semicontinuity of I, namely

$$u_\nu \rightharpoonup \overline{u} \text{ in } W^{1,p} \implies \liminf_{\nu \to \infty} I(u_\nu) \geq I(\overline{u}).$$

Since $\{u_\nu\}$ is a minimizing sequence, we deduce that $\overline{u}$ is a minimizer of (P).

In Section 3.4 we derive the *Euler-Lagrange equation* associated to (P). Since the solution of (P) is only in a Sobolev space, we are able to write only a weak form of this equation.

In Section 3.5 we say some words about the considerably harder case where the unknown function u is a vector, i.e. $u : \Omega \subset \mathbb{R}^n \to \mathbb{R}^N$, with $n, N > 1$.

In Section 3.6 we briefly explain what can be done, in some cases, when the hypothesis (H_1) of convexity fails to hold.

The interested reader is referred for further reading to the book of the author [31] or to Buttazzo [15], Buttazzo-Giaquinta-Hildebrandt [17], Cesari [20], Ekeland-Temam [42], Giaquinta [49], Giusti [53], Ioffe-Tihomirov [66], Morrey [79], Struwe [97] and Zeidler [104].

Let $u_\nu \in u_0 + W_0^{1,2}(\Omega)$ be a minimizing sequence of (D), this means that

$$I(u_\nu) \to \inf \{I(u)\} = m, \text{ as } \nu \to \infty.$$

Observe that by Poincaré inequality (cf. Theorem 1.48) we can find constants $\gamma_1, \gamma_2 > 0$ so that

$$\sqrt{2I(u_\nu)} = \|\nabla u_\nu\|_{L^2} \geq \gamma_1 \|u_\nu\|_{W^{1,2}} - \gamma_2 \|u_0\|_{W^{1,2}}.$$

Since u_ν is a minimizing sequence and $m < \infty$ we deduce that there exists $\gamma_3 > 0$ so that

$$\|u_\nu\|_{W^{1,2}} \leq \gamma_3.$$

Applying Exercise 1.4.5 we deduce that there exists $\overline{u} \in u_0 + W_0^{1,2}(\Omega)$ and a subsequence (still denoted u_ν) so that

$$u_\nu \rightharpoonup \overline{u} \text{ in } W^{1,2}, \text{ as } \nu \to \infty.$$

Step 2 (Lower semicontinuity). We now show that I is (sequentially) weakly lower semicontinuous; this means that

$$u_\nu \rightharpoonup \overline{u} \text{ in } W^{1,2} \Rightarrow \liminf_{\nu \to \infty} I(u_\nu) \geq I(\overline{u}).$$

This step is independent of the fact that $\{u_\nu\}$ is a minimizing sequence. We trivially have that

$$|\nabla u_\nu|^2 = |\nabla \overline{u}|^2 + 2 \langle \nabla \overline{u}; \nabla u_\nu - \nabla \overline{u} \rangle + |\nabla u_\nu - \nabla \overline{u}|^2$$
$$\geq |\nabla \overline{u}|^2 + 2 \langle \nabla \overline{u}; \nabla u_\nu - \nabla \overline{u} \rangle.$$

Integrating this expression we have

$$I(u_\nu) \geq I(\overline{u}) + \int_\Omega \langle \nabla \overline{u}; \nabla u_\nu - \nabla \overline{u} \rangle \, dx.$$

To conclude we show that the second term on the right-hand side of the inequality tends to 0. Indeed since $\nabla \overline{u} \in L^2$ and $\nabla u_\nu - \nabla \overline{u} \rightharpoonup 0$ in L^2 this implies, by definition of weak convergence in L^2, that

$$\lim_{\nu \to \infty} \int_\Omega \langle \nabla \overline{u}; \nabla u_\nu - \nabla \overline{u} \rangle \, dx = 0.$$

Therefore returning to the above inequality we have indeed obtained that

$$\liminf_{\nu \to \infty} I(u_\nu) \geq I(\overline{u}).$$

Part 4 (Converse). We finally prove that if $\overline{u} \in u_0 + W_0^{1,2}(\Omega)$ satisfies (3.1) then it is necessarily a minimizer of (D). Let $u \in u_0 + W_0^{1,2}(\Omega)$ be any element and set $\varphi = u - \overline{u}$. Observe that $\varphi \in W_0^{1,2}(\Omega)$ and

$$I(u) = I(\overline{u} + \varphi) = \int_{\Omega} \frac{1}{2} |\nabla \overline{u} + \nabla \varphi|^2 \, dx$$

$$= I(\overline{u}) + \int_{\Omega} \langle \nabla \overline{u}; \nabla \varphi \rangle \, dx + I(\varphi) \geq I(\overline{u})$$

since the second term is 0 according to (3.1) and the last one is non-negative. This achieves the proof of the theorem. ∎

3.2.1   Exercise

Exercise 3.2.1 Let Ω be as in the theorem and $h \in L^2(\Omega)$. Show that

$$(P) \quad \inf \left\{ I(u) = \int_{\Omega} \left[\frac{1}{2} |\nabla u(x)|^2 - h(x) u(x) \right] dx : u \in W_0^{1,2}(\Omega) \right\} = m$$

has a unique solution $\overline{u} \in W_0^{1,2}(\Omega)$ which satisfies in addition

$$\int_{\Omega} \langle \nabla \overline{u}(x); \nabla \varphi(x) \rangle \, dx = \int_{\Omega} h(x) \varphi(x) \, dx, \ \forall \varphi \in W_0^{1,2}(\Omega).$$

3.3   A general existence theorem

The main theorem of the present chapter is the following.

Theorem 3.3 *Let $\Omega \subset \mathbb{R}^n$ be a bounded open set with Lipschitz boundary. Let $f \in C^0(\overline{\Omega} \times \mathbb{R} \times \mathbb{R}^n)$, $f = f(x, u, \xi)$, satisfy*

(H_1) $\xi \to f(x, u, \xi)$ *is convex for every $(x, u) \in \overline{\Omega} \times \mathbb{R}$;*

(H_2) *there exist $p > q \geq 1$ and $\alpha_1 > 0$, $\alpha_2, \alpha_3 \in \mathbb{R}$ such that*

$$f(x, u, \xi) \geq \alpha_1 |\xi|^p + \alpha_2 |u|^q + \alpha_3, \ \forall (x, u, \xi) \in \overline{\Omega} \times \mathbb{R} \times \mathbb{R}^n.$$

Let

$$(P) \quad \inf \left\{ I(u) = \int_{\Omega} f(x, u(x), \nabla u(x)) \, dx : u \in u_0 + W_0^{1,p}(\Omega) \right\} = m$$

where $u_0 \in W^{1,p}(\Omega)$ with $I(u_0) < \infty$. Then there exists $\overline{u} \in u_0 + W_0^{1,p}(\Omega)$ a minimizer of (P).

Furthermore if $(u, \xi) \to f(x, u, \xi)$ is strictly convex for every $x \in \overline{\Omega}$, then the minimizer is unique.

that satisfies all the hypotheses of the theorem but (H_2), this hypothesis is only verified with $p = 1$. We will see in Chapter 5 that this failure may lead to non-existence of minima for the corresponding (P). The reason why $p = 1$ is not allowed is that the corresponding Sobolev space $W^{1,1}$ is not reflexive (see Chapter 1).

Example 3.7 This example is of the minimal surface type but easier, it also shows that all the hypotheses of the theorem are satisfied, except (H_2) that is true with $p = 1$. This weakening of (H_2) leads to the following counterexample. Let $n = 1$,

$$f(x, u, \xi) = f(u, \xi) = \sqrt{u^2 + \xi^2}$$

$$(P) \quad \inf \left\{ I(u) = \int_0^1 f(u(x), u'(x)) \, dx : u \in X \right\} = m$$

where $X = \{u \in W^{1,1}(0, 1) : u(0) = 0, \ u(1) = 1\}$. Let us prove that (P) has no solution. We first show that $m = 1$ and start by observing that $m \geq 1$ since

$$I(u) \geq \int_0^1 |u'(x)| \, dx \geq \int_0^1 u'(x) \, dx = u(1) - u(0) = 1.$$

To establish that $m = 1$, we construct a minimizing sequence $u_\nu \in X$ (ν being an integer) as follows

$$u_\nu(x) = \begin{cases} 0 & \text{if } x \in \left[0, 1 - \frac{1}{\nu}\right] \\ 1 + \nu(x - 1) & \text{if } x \in \left(1 - \frac{1}{\nu}, 1\right]. \end{cases}$$

We therefore have $m = 1$ since

$$1 \leq I(u_\nu) = \int_{1-\frac{1}{\nu}}^1 \sqrt{(1 + \nu(x-1))^2 + \nu^2} \, dx$$

$$\leq \frac{1}{\nu}\sqrt{1 + \nu^2} \to 1, \text{ as } \nu \to \infty.$$

Assume now, for the sake of contradiction, that there exists $\overline{u} \in X$ a minimizer of (P). We should then have, as above,

$$1 = I(\overline{u}) = \int_0^1 \sqrt{\overline{u}^2 + \overline{u}'^2} \, dx \geq \int_0^1 |\overline{u}'| \, dx$$

$$\geq \int_0^1 \overline{u}' \, dx = \overline{u}(1) - \overline{u}(0) = 1.$$

This implies that $\overline{u} = 0$ a.e. in $(0, 1)$. Since elements of X are continuous we have that $\overline{u} \equiv 0$ and this is incompatible with the boundary data. Thus (P) has no solution.

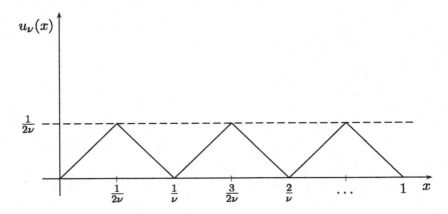

Figure 3.1: minimizing sequence

a minimizer of (P), i.e. $I(\overline{u}) = 0$. This implies that $\overline{u} = 0$ and $|\overline{u}'| = 1$ a.e. in $(0, 1)$. Since the elements of $W^{1,4}$ are continuous we have that $\overline{u} \equiv 0$ and hence $\overline{u}' \equiv 0$ which is clearly absurd.

So let us show that $m = 0$ by constructing an appropriate minimizing sequence. Let $u_\nu \in W_0^{1,4}$ ($\nu \geq 2$ being an integer) defined on each interval $[k/\nu, (k+1)/\nu]$, $0 \leq k \leq \nu - 1$, as follows

$$u_\nu(x) = \begin{cases} x - \frac{k}{\nu} & \text{if } x \in \left[\frac{2k}{2\nu}, \frac{2k+1}{2\nu}\right] \\ -x + \frac{k+1}{\nu} & \text{if } x \in \left(\frac{2k+1}{2\nu}, \frac{2k+2}{2\nu}\right]. \end{cases}$$

Observe that $|u_\nu'| = 1$ a.e. and $|u_\nu| \leq 1/(2\nu)$ leading therefore to the desired convergence, namely

$$0 \leq I(u_\nu) \leq \frac{1}{(2\nu)^4} \to 0, \text{ as } \nu \to \infty.$$

Proof. We do not prove the theorem in its full generality. We refer to the literature and in particular to Corollary 3.24 in [31, 2nd edition] for a general proof; see also the exercises below. We prove it under the following stronger hypotheses. We assume that $f \in C^1(\overline{\Omega} \times \mathbb{R} \times \mathbb{R}^n)$, instead of C^0, and

(H_{1+}) $(u, \xi) \to f(x, u, \xi)$ is convex for every $x \in \overline{\Omega}$;

(H_{2+}) there exist $p > 1$ and $\alpha_1 > 0$, $\alpha_3 \in \mathbb{R}$ such that

$$f(x, u, \xi) \geq \alpha_1 |\xi|^p + \alpha_3, \ \forall (x, u, \xi) \in \overline{\Omega} \times \mathbb{R} \times \mathbb{R}^n.$$

This step is independent of the fact that $\{u_\nu\}$ is a minimizing sequence. Using the convexity of f and the fact that it is C^1 we get

$$f(x, u_\nu, \nabla u_\nu) \geq$$
$$f(x, \overline{u}, \nabla \overline{u}) + f_u(x, \overline{u}, \nabla \overline{u})(u_\nu - \overline{u}) + \langle f_\xi(x, \overline{u}, \nabla \overline{u}) ; \nabla u_\nu - \nabla \overline{u} \rangle. \qquad (3.3)$$

Before proceeding further we need to show that the combination of (H_3) and $\overline{u} \in W^{1,p}(\Omega)$ leads to

$$f_u(x, \overline{u}, \nabla \overline{u}) \in L^{p'}(\Omega) \quad \text{and} \quad f_\xi(x, \overline{u}, \nabla \overline{u}) \in L^{p'}(\Omega; \mathbb{R}^n) \qquad (3.4)$$

where $1/p + 1/p' = 1$ (i.e. $p' = p/(p-1)$). Indeed let us prove the first statement, the other one being shown similarly. We have (β_1 being a constant)

$$\int_\Omega |f_u(x, \overline{u}, \nabla \overline{u})|^{p'} \, dx \leq \beta^{p'} \int_\Omega \left(1 + |\overline{u}|^{p-1} + |\nabla \overline{u}|^{p-1}\right)^{\frac{p}{p-1}} dx$$
$$\leq \beta_1 \left(1 + \|\overline{u}\|^p_{W^{1,p}}\right) < \infty.$$

Using Hölder inequality and (3.4) we find that for $u_\nu \in W^{1,p}(\Omega)$

$$f_u(x, \overline{u}, \nabla \overline{u})(u_\nu - \overline{u}), \ \langle f_\xi(x, \overline{u}, \nabla \overline{u}) ; \nabla u_\nu - \nabla \overline{u} \rangle \in L^1(\Omega).$$

We next integrate (3.3) to get

$$I(u_\nu) \geq I(\overline{u}) + \int_\Omega [f_u(x, \overline{u}, \nabla \overline{u})(u_\nu - \overline{u}) + \langle f_\xi(x, \overline{u}, \nabla \overline{u}) ; \nabla u_\nu - \nabla \overline{u} \rangle] \, dx \quad (3.5)$$

Since $u_\nu - \overline{u} \rightharpoonup 0$ in $W^{1,p}$ (i.e. $u_\nu - \overline{u} \rightharpoonup 0$ in L^p and $\nabla u_\nu - \nabla \overline{u} \rightharpoonup 0$ in L^p) and (3.4) holds, we deduce, from the definition of weak convergence in L^p, that

$$\lim_{\nu \to \infty} \int_\Omega f_u(x, \overline{u}, \nabla \overline{u})(u_\nu - \overline{u}) \, dx = \lim_{\nu \to \infty} \int_\Omega \langle f_\xi(x, \overline{u}, \nabla \overline{u}) ; \nabla u_\nu - \nabla \overline{u} \rangle \, dx = 0.$$

Therefore returning to (3.5) we have indeed obtained that

$$\liminf_{\nu \to \infty} I(u_\nu) \geq I(\overline{u}).$$

Step 3. We now combine the two steps. Since $\{u_\nu\}$ is a minimizing sequence (i.e. $I(u_\nu) \to \inf\{I(u)\} = m$) and for such a sequence we have lower semicontinuity (i.e. $\liminf I(u_\nu) \geq I(\overline{u})$), we deduce that $I(\overline{u}) = m$, i.e. $\overline{u}$ is a minimizer of (P).

Part 2 (Uniqueness). The proof is almost identical to the one of Theorem 3.1 and Theorem 2.1. Assume that there exist $\overline{u}, \overline{v} \in u_0 + W_0^{1,p}(\Omega)$ so that

$$I(\overline{u}) = I(\overline{v}) = m$$

Case 1: $p > n$. For every $R > 0$, there exists $\gamma = \gamma(R)$ such that

$$|g(x, u) - g(x, v)| \leq \gamma |u - v|$$

for every $x \in \overline{\Omega}$ and every $u, v \in \mathbb{R}$ with $|u|, |v| \leq R$.

Case 2: $p = n$. There exist $q \geq 1$ and $\gamma > 0$ so that

$$|g(x, u) - g(x, v)| \leq \gamma \left(1 + |u|^{q-1} + |v|^{q-1}\right) |u - v|$$

for every $x \in \overline{\Omega}$ and every $u, v \in \mathbb{R}$.

Case 3: $p < n$. There exist $q \in [1, np/(n-p))$ and $\gamma > 0$ so that

$$|g(x, u) - g(x, v)| \leq \gamma \left(1 + |u|^{q-1} + |v|^{q-1}\right) |u - v|$$

for every $x \in \overline{\Omega}$ and every $u, v \in \mathbb{R}$.

Exercise 3.3.3 Prove Theorem 3.3 in the following framework. Let $\alpha, \beta \in \mathbb{R}^N$, $N > 1$ and

$$(P) \quad \inf_{u \in X} \left\{ I(u) = \int_a^b f(x, u(x), u'(x)) \, dx \right\} = m$$

where $X = \{u \in W^{1,p}((a,b); \mathbb{R}^N) : u(a) = \alpha, \ u(b) = \beta\}$ and

(i) $f \in C^1\left([a,b] \times \mathbb{R}^N \times \mathbb{R}^N\right)$, $(u, \xi) \to f(x, u, \xi)$ is convex for every $x \in [a, b]$;

(ii) there exist $p > q \geq 1$ and $\alpha_1 > 0$, $\alpha_2, \alpha_3 \in \mathbb{R}$ such that

$$f(x, u, \xi) \geq \alpha_1 |\xi|^p + \alpha_2 |u|^q + \alpha_3, \ \forall (x, u, \xi) \in [a, b] \times \mathbb{R}^N \times \mathbb{R}^N;$$

(iii) for every $R > 0$, there exists $\beta = \beta(R)$ such that

$$|f_u(x, u, \xi)| \leq \beta(1 + |\xi|^p) \quad \text{and} \quad |f_\xi(x, u, \xi)| \leq \beta\left(1 + |\xi|^{p-1}\right)$$

for every $x \in [a, b]$ and every $u, \xi \in \mathbb{R}^N$ with $|u| \leq R$.

Exercise 3.3.4 Let

$$f(x, u, \xi) = g(x, u) + h(x, \xi)$$

with $u \to g(x, u)$, $\xi \to h(x, \xi)$ convex and at least one of them strictly convex. Prove that the uniqueness in Theorem 3.3 is conserved.

It can be weakened, but only slightly by the use of Sobolev imbedding theorem (see Exercise 3.4.1), or by requiring only measurability in x. In this last case one can assume that f is a Carathéodory function (cf. Remark 3.4 (i)) that satisfies for almost every $x \in \Omega$ and every $(u, \xi) \in \mathbb{R} \times \mathbb{R}^n$

$$|f_u(x, u, \xi)|, \ |f_\xi(x, u, \xi)| \leq \beta_1(x) + \beta_2 \left(|u|^{p-1} + |\xi|^{p-1} \right)$$

where $\beta_1 \in L^{p'}(\Omega)$ and $\beta_2 \geq 0$.

(iii) Of course any solution of (E) is a solution of (E_w). The converse is true only if $\overline{u}$ is sufficiently regular.

(iv) In the statement of the theorem we do not need hypothesis (H_1) or (H_2) of Theorem 3.3. Therefore we do not use the convexity of f (naturally for the converse we need the convexity of f). However we require that a minimizer of (P) does exist.

(v) The theorem remains valid in the vectorial case, where $u : \Omega \subset \mathbb{R}^n \to \mathbb{R}^N$, with $n, N > 1$. The Euler-Lagrange equation becomes now a system of partial differential equations and reads as follows

$$(E) \quad \sum_{i=1}^n \frac{\partial}{\partial x_i} \left[f_{\xi_i^j}(x, \overline{u}, \nabla \overline{u}) \right] = f_{u^j}(x, \overline{u}, \nabla \overline{u}), \ \forall x \in \overline{\Omega}, \ j = 1, \cdots, N$$

where $f : \overline{\Omega} \times \mathbb{R}^N \times \mathbb{R}^{N \times n} \to \mathbb{R}$ and

$$u = \left(u^1, \cdots, u^N \right) \in \mathbb{R}^N, \ \xi = \left(\xi_i^j \right)_{1 \leq i \leq n}^{1 \leq j \leq N} \in \mathbb{R}^{N \times n} \text{ and } \nabla u = \left(\frac{\partial u^j}{\partial x_i} \right)_{1 \leq i \leq n}^{1 \leq j \leq N}.$$

(vi) In some cases one can be interested in an even weaker form of the Euler-Lagrange equation. More precisely if we choose the test functions φ in (E_w) to be $C_0^\infty(\Omega)$ instead of $W_0^{1,p}(\Omega)$, then one can weaken the hypothesis (H_3) and replace it by

(H_3') there exist $p \geq 1$ and $\beta \geq 0$ so that for every $(x, u, \xi) \in \overline{\Omega} \times \mathbb{R} \times \mathbb{R}^n$

$$|f_u(x, u, \xi)|, \ |f_\xi(x, u, \xi)| \leq \beta(1 + |u|^p + |\xi|^p).$$

The proof of the theorem remains almost identical. The choice of the space, where the test function φ belongs, depends on the context. If we want to use the solution, $\overline{u}$, itself as a test function then we are obliged to choose $W_0^{1,p}(\Omega)$ as the right space (see Sections 4.3 and 4.4) while some other times (see Section 4.2) we can actually limit ourselves to the space $C_0^\infty(\Omega)$.

Proof. The proof is divided into four steps.

Applying Lebesgue dominated convergence theorem we deduce that (3.7) holds.

Step 3 (Derivation of (E_w) and (E)). The conclusion of the theorem follows from the preceding step. Indeed since $\overline{u}$ is a minimizer of (P) then

$$I\left(\overline{u} + \epsilon\varphi\right) \geq I\left(\overline{u}\right), \ \forall\varphi \in W_0^{1,p}\left(\Omega\right)$$

and thus

$$\lim_{\epsilon\to 0}\frac{I\left(\overline{u} + \epsilon\varphi\right) - I\left(\overline{u}\right)}{\epsilon} = 0$$

which combined with (3.7) implies (E_w).

To get (E) it remains to integrate by parts (using Exercise 1.4.8) and to find

$$(E_w) \ \int_\Omega \left[f_u\left(x, \overline{u}, \nabla\overline{u}\right) - \operatorname{div}\left[f_\xi\left(x, \overline{u}, \nabla\overline{u}\right)\right]\right]\varphi\,dx = 0, \ \forall\varphi \in W_0^{1,p}\left(\Omega\right).$$

The fundamental lemma of the calculus of variations (Theorem 1.24) implies the claim.

Step 4 (Converse). Let $\overline{u}$ be a solution of (E_w) (note that any solution of (E) is necessarily a solution of (E_w)). From the convexity of f we deduce that for every $u \in u_0 + W_0^{1,p}\left(\Omega\right)$ the following holds

$$\begin{aligned}
f\left(x, u, \nabla u\right) &\geq f\left(x, \overline{u}, \nabla\overline{u}\right) + f_u\left(x, \overline{u}, \nabla\overline{u}\right)\left(u - \overline{u}\right) \\
&\quad + \left\langle f_\xi\left(x, \overline{u}, \nabla\overline{u}\right); \left(\nabla u - \nabla\overline{u}\right)\right\rangle.
\end{aligned}$$

Integrating, using (E_w) and the fact that $u - \overline{u} \in W_0^{1,p}\left(\Omega\right)$ we immediately get that $I\left(u\right) \geq I\left(\overline{u}\right)$ and hence the theorem. ∎

We now discuss some examples.

Example 3.13 In the case of Dirichlet integral we have

$$f\left(x, u, \xi\right) = f\left(\xi\right) = \frac{1}{2}\left|\xi\right|^2$$

which satisfies (H_3). The equation (E_w) is then

$$\int_\Omega \left\langle\nabla\overline{u}\left(x\right); \nabla\varphi\left(x\right)\right\rangle dx = 0, \ \forall\varphi \in W_0^{1,2}\left(\Omega\right)$$

while (E) is $\Delta\overline{u} = 0$.

Example 3.14 Consider the generalization of the preceding example, where

$$f\left(x, u, \xi\right) = f\left(\xi\right) = \frac{1}{p}\left|\xi\right|^p.$$

Example 3.17 (Poincaré-Wirtinger inequality) Let $\lambda > \pi$, $n = 1$ and

$$f(x, u, \xi) = f(u, \xi) = \frac{1}{2} \left(\xi^2 - \lambda^2 u^2 \right)$$

$$(P) \quad \inf \left\{ I(u) = \int_0^1 f(u(x), u'(x)) \, dx : u \in W_0^{1,2}(0, 1) \right\} = m.$$

Note that $\xi \to f(u, \xi)$ is convex while $(u, \xi) \to f(u, \xi)$ is not. We have seen that $m = -\infty$ and therefore (P) has no minimizer; however the Euler-Lagrange equation

$$u'' + \lambda^2 u = 0 \text{ in } [0, 1]$$

has $u \equiv 0$ as a solution. It is therefore not a minimizer.

Example 3.18 Let $n = 1$,

$$f(x, u, \xi) = f(\xi) = \left(\xi^2 - 1 \right)^2$$

which is non-convex, and

$$(P) \quad \inf \left\{ I(u) = \int_0^1 f(u'(x)) \, dx : u \in W_0^{1,4}(0, 1) \right\} = m.$$

We have seen that $m = 0$. The Euler-Lagrange equation is

$$(E) \quad \frac{d}{dx} \left[\overline{u}' \left(\overline{u}'^2 - 1 \right) \right] = 0$$

and its weak form is (note that f satisfies (H_3))

$$(E_w) \quad \int_0^1 \overline{u}' \left(\overline{u}'^2 - 1 \right) \varphi' \, dx = 0, \ \forall \varphi \in W_0^{1,4}(0, 1).$$

It is clear that $\overline{u} \equiv 0$ is a solution of (E) and (E_w), but it is not a minimizer of (P) since $m = 0$ and $I(0) = 1$. The present example is also interesting for another reason. Indeed the function

$$v(x) = \begin{cases} x & \text{if } x \in [0, 1/2] \\ 1 - x & \text{if } x \in (1/2, 1] \end{cases}$$

is clearly a minimizer of (P) which is not C^1; it satisfies (E_w) but not (E).

- $u = \left(u^1, \cdots, u^N\right) \in \mathbb{R}^N,\ \xi = \left(\xi_i^j\right)_{1 \le i \le n}^{1 \le j \le N} \in \mathbb{R}^{N \times n}$ and $\nabla u = \left(\frac{\partial u^j}{\partial x_i}\right)_{1 \le i \le n}^{1 \le j \le N}$;

- $u \in u_0 + W_0^{1,p}\left(\Omega; \mathbb{R}^N\right)$ means that $u^j, u_0^j \in W^{1,p}\left(\Omega\right),\ j = 1, \cdots, N$, and $u - u_0 \in W_0^{1,p}\left(\Omega; \mathbb{R}^N\right)$ (which roughly means that $u = u_0$ on $\partial\Omega$).

All the results of the preceding sections apply to the present context when $n, N > 1$. However, while for $N = 1$ (or analogously when $n = 1$) Theorem 3.3 is almost optimal, it is now far from being so. The vectorial case is intrinsically more difficult. For example the Euler-Lagrange equations associated to (P) are then a system of partial differential equations, whose treatment is considerably harder than that of a single partial differential equation.

We present one extension of Theorem 3.3; it is not the best possible result, but it has the advantage of giving some flavours of what can be done. For the sake of clarity we essentially consider only the case $n = N = 2$; but, in a remark, we briefly mention what can be done in the higher dimensional case.

Theorem 3.19 *Let $n = N = 2$ and $\Omega \subset \mathbb{R}^2$ be a bounded open set with Lipschitz boundary. Let*

$$f : \overline{\Omega} \times \mathbb{R}^2 \times \mathbb{R}^{2\times 2} \to \mathbb{R},\ f = f\left(x, u, \xi\right),$$

$$F : \overline{\Omega} \times \mathbb{R}^2 \times \mathbb{R}^{2\times 2} \times \mathbb{R} \to \mathbb{R},\ F = F\left(x, u, \xi, \delta\right),$$

be continuous and satisfying

$$f\left(x, u, \xi\right) = F\left(x, u, \xi, \det \xi\right),\ \forall \left(x, u, \xi\right) \in \overline{\Omega} \times \mathbb{R}^2 \times \mathbb{R}^{2\times 2}$$

where $\det \xi$ denotes the determinant of the matrix ξ. Assume also that

(H_1^{vect}) *$(\xi, \delta) \to F\left(x, u, \xi, \delta\right)$ is convex for every $(x, u) \in \overline{\Omega} \times \mathbb{R}^2$;*

(H_2^{vect}) *there exist $p > \max\left[q, 2\right]$ and $\alpha_1 > 0$, $\alpha_2, \alpha_3 \in \mathbb{R}$ such that*

$$F\left(x, u, \xi, \delta\right) \ge \alpha_1 \left|\xi\right|^p + \alpha_2 \left|u\right|^q + \alpha_3,\ \forall \left(x, u, \xi, \delta\right) \in \overline{\Omega} \times \mathbb{R}^2 \times \mathbb{R}^{2\times 2} \times \mathbb{R}.$$

Let $u_0 \in W^{1,p}\left(\Omega; \mathbb{R}^2\right)$ be such that $I\left(u_0\right) < \infty$, then (P) has at least one solution.

Remark 3.20 (i) It is clear that, from the point of view of convexity, the theorem is more general than Theorem 3.3. Indeed if $\xi \to f\left(x, u, \xi\right)$ is convex then choose $F\left(x, u, \xi, \delta\right) = f\left(x, u, \xi\right)$ and therefore (H_1^{vect}) and (H_1) are equivalent. However (H_1^{vect}) is more general since, for example, a function of the form

$$f\left(x, u, \xi\right) = \left|\xi\right|^4 + 16 \left(\det \xi\right)^2$$

is non-convex (cf. Exercise 3.5.1), while

$$F\left(x, u, \xi, \delta\right) = \left|\xi\right|^4 + 16\delta^2$$

Let us now see two examples.

Example 3.21 Let $n = N = 2$, $p > 2$ and

$$f(x, u, \xi) = f(\xi) = \frac{1}{p} |\xi|^p + h(\det \xi)$$

where $h : \mathbb{R} \to \mathbb{R}$ is non-negative and convex (for example $h(\det \xi) = (\det \xi)^2$). All hypotheses of the theorem are clearly satisfied. It is also interesting to compute the associated Euler-Lagrange equation. To make them simple consider only the case $p = 2$ and set

$$u = u(x_1, x_2) = \left(u^1(x_1, x_2), u^2(x_1, x_2)\right).$$

The system is then

$$\begin{cases} \Delta u^1 + \left[h'(\det \nabla u) u_{x_2}^2\right]_{x_1} - \left[h'(\det \nabla u) u_{x_1}^2\right]_{x_2} = 0 \\ \Delta u^2 - \left[h'(\det \nabla u) u_{x_2}^1\right]_{x_1} + \left[h'(\det \nabla u) u_{x_1}^1\right]_{x_2} = 0. \end{cases}$$

Example 3.22 Another important example coming from applications is the following. Let $n = N = 3$, $p > 3$, $q \geq 1$ and

$$f(x, u, \xi) = f(\xi) = \alpha |\xi|^p + \beta |\mathrm{adj}_2 \xi|^q + h(\det \xi)$$

where $h : \mathbb{R} \to \mathbb{R}$ is non-negative and convex and $\alpha, \beta > 0$.

The key ingredient in the proof of the theorem is the following lemma that is due to Morrey and Reshetnyak.

Lemma 3.23 *Let $\Omega \subset \mathbb{R}^2$ be a bounded open set with Lipschitz boundary, $p > 2$ and*

$$u^\nu = (\varphi^\nu, \psi^\nu) \rightharpoonup u = (\varphi, \psi) \ in \ W^{1,p}(\Omega; \mathbb{R}^2);$$

then

$$\det \nabla u^\nu \rightharpoonup \det \nabla u \ in \ L^{p/2}(\Omega).$$

Remark 3.24 **(i)** At first glance the result is a little surprising. Indeed we have seen in Chapter 1 (in particular Exercise 1.3.3) that if two sequences, say $(\varphi^\nu)_{x_1}$ and $(\psi^\nu)_{x_2}$, converge weakly respectively to φ_{x_1} and ψ_{x_2}, then, in general, their product $(\varphi^\nu)_{x_1} (\psi^\nu)_{x_2}$ does not converge weakly to $\varphi_{x_1} \psi_{x_2}$. Writing

$$\det \nabla u^\nu = (\varphi^\nu)_{x_1} (\psi^\nu)_{x_2} - (\varphi^\nu)_{x_2} (\psi^\nu)_{x_1}$$

we see that both terms $(\varphi^\nu)_{x_1} (\psi^\nu)_{x_2}$ and $(\varphi^\nu)_{x_2} (\psi^\nu)_{x_1}$ do not, in general, converge weakly to $\varphi_{x_1} \psi_{x_2}$ and $\varphi_{x_2} \psi_{x_1}$ but, according to the lemma, their difference, which is $\det \nabla u^\nu$, converges weakly to their difference, namely $\det \nabla u$. We

The result (3.8) then easily follows, see below. In the sequel we write

$$u^\nu = (\varphi^\nu, \psi^\nu) \quad \text{and} \quad u = (\varphi, \psi).$$

Indeed from Rellich theorem (Theorem 1.44) we have, since $\varphi^\nu \rightharpoonup \varphi$ in $W^{1,p}$ and $p > 2$, that $\varphi^\nu \to \varphi$ in L^∞. Combining this observation with the fact that

$$\psi^\nu_{x_1}, \psi^\nu_{x_2} \rightharpoonup \psi_{x_1}, \psi_{x_2} \text{ in } L^p$$

we deduce (cf. Exercise 1.3.3) that

$$\varphi^\nu \psi^\nu_{x_1}, \varphi^\nu \psi^\nu_{x_2} \rightharpoonup \varphi \psi_{x_1}, \varphi \psi_{x_2} \text{ in } L^p. \tag{3.11}$$

Since $v_{x_1}, v_{x_2} \in C_0^\infty \subset L^{p'}$, we deduce from (3.9), applied to $w = u^\nu$, and from (3.11) that

$$\lim_{\nu \to \infty} \iint_\Omega \det \nabla u^\nu \, v \, dx_1 dx_2 = - \iint_\Omega [\varphi \psi_{x_2} v_{x_1} - \varphi \psi_{x_1} v_{x_2}] \, dx_1 dx_2.$$

Using again (3.9), applied to $w = u$, we have indeed obtained the claimed result (3.8).

Step 2. We now show that (3.8) still holds under the further hypothesis $v \in C_0^\infty(\Omega)$, but considering now the general case, i.e. $u^\nu, u \in W^{1,p}(\Omega; \mathbb{R}^2)$.

In fact (3.9) continues to hold under the weaker hypothesis that $v \in C_0^\infty(\Omega)$ and $w \in W^{1,p}(\Omega; \mathbb{R}^2)$; of course the proof must be different, since this time we only know that $w \in W^{1,p}(\Omega; \mathbb{R}^2)$. Let us postpone for a moment the proof of this fact and observe that if (3.9) holds for $w \in W^{1,p}(\Omega; \mathbb{R}^2)$ then, with exactly the same argument as in the previous step, we get (3.8) under the hypotheses $v \in C_0^\infty(\Omega)$ and $u^\nu, u \in W^{1,p}(\Omega; \mathbb{R}^2)$.

We now prove the above claim and we start by regularizing $w \in W^{1,p}(\Omega; \mathbb{R}^2)$ appealing to Theorem 1.34. We therefore find for every $\epsilon > 0$, a function $w^\epsilon = (\varphi^\epsilon, \psi^\epsilon) \in C^2(\overline{\Omega}; \mathbb{R}^2)$ so that

$$\|w - w^\epsilon\|_{W^{1,p}} \leq \epsilon \quad \text{and} \quad \|w - w^\epsilon\|_{L^\infty} \leq \epsilon.$$

Since $p \geq 2$ we can find (cf. Exercise 3.5.4) a constant α_1 (independent of ϵ but depending on w) so that

$$\|\det \nabla w - \det \nabla w^\epsilon\|_{L^{p/2}} \leq \alpha_1 \epsilon. \tag{3.12}$$

It is also easy to see that we have, for α_2 a constant (independent of ϵ but depending on w),

$$\|\varphi \psi_{x_2} - \varphi^\epsilon \psi^\epsilon_{x_2}\|_{L^p} \leq \alpha_2 \epsilon, \quad \|\varphi \psi_{x_1} - \varphi^\epsilon \psi^\epsilon_{x_1}\|_{L^p} \leq \alpha_2 \epsilon \tag{3.13}$$

The previous step has shown that

$$\lim_{\nu \to \infty} \left| \iint_\Omega \left(\det \nabla u^\nu - \det \nabla u \right) v^\epsilon \right| = 0$$

while (3.14), the fact that $u^\nu \rightharpoonup u$ in $W^{1,p}$ and Exercise 3.5.4 show that we can find $\gamma > 0$ so that

$$\| v - v^\epsilon \|_{L^{p/(p-2)}} \| \det \nabla u^\nu - \det \nabla u \|_{L^{p/2}} \leq \gamma \epsilon.$$

Since ϵ is arbitrary we have indeed obtained that (3.8) holds for $v \in L^{p/(p-2)}$ and for $u^\nu \rightharpoonup u$ in $W^{1,p}$. The lemma is therefore proved. ∎

We can now proceed with the proof of Theorem 3.19.

Proof. We prove the theorem under the further hypotheses (for a general proof see Theorem 8.31 in [31, 2nd edition])

$$f(x, u, \xi) = g(x, u, \xi) + h(x, \det \xi)$$

where g satisfies (H_1) and (H_2), with $p > 2$, of Theorem 3.3 and $h \in C^1\left(\overline{\Omega} \times \mathbb{R}\right)$, $h \geq 0$, $\delta \to h(x, \delta)$ is convex for every $x \in \overline{\Omega}$ and there exists $\gamma > 0$ so that

$$|h_\delta(x, \delta)| \leq \gamma \left(1 + |\delta|^{(p-2)/2} \right). \tag{3.15}$$

The proof is then identical to the one of Theorem 3.3, except the second step (the weak lower semicontinuity), that we discuss now. We have to prove that

$$u_\nu \rightharpoonup \overline{u} \text{ in } W^{1,p} \ \Rightarrow \ \liminf_{\nu \to \infty} I(u_\nu) \geq I(\overline{u})$$

where $I(u) = G(u) + H(u)$ with

$$G(u) = \int_\Omega g(x, u(x), \nabla u(x))\, dx, \quad H(u) = \int_\Omega h(x, \det \nabla u(x))\, dx.$$

We have already proved in Theorem 3.3 that

$$\liminf_{\nu \to \infty} G(u_\nu) \geq G(\overline{u})$$

and therefore the result follows if we can show

$$\liminf_{\nu \to \infty} H(u_\nu) \geq H(\overline{u}).$$

Since h is convex and C^1 we have

$$h(x, \det \nabla u_\nu) \geq h(x, \det \nabla \overline{u}) + h_\delta(x, \det \nabla \overline{u})(\det \nabla u_\nu - \det \nabla \overline{u}). \tag{3.16}$$

Remark 3.26 (i) If Ω is bounded we then have the following relations

$$u_\nu \overset{*}{\rightharpoonup} u \text{ in } L^\infty \;\Rightarrow\; u_\nu \rightharpoonup u \text{ in } L^1 \;\Rightarrow\; u_\nu \rightharpoonup u \text{ in } \mathcal{D}'.$$

(ii) The definition can be generalized to u_ν and u that are not necessarily in $L^1_{\mathrm{loc}}(\Omega)$, but are merely what is known as "distributions", cf. Exercise 3.5.6.

Exercise 3.5.1 Let $\xi \in \mathbb{R}^{2\times2}$. Prove that the functions

$$f_1(\xi) = (\det \xi)^2 \quad \text{and} \quad f_2(\xi) = |\xi|^4 + 16(\det \xi)^2$$

are not convex.

Exercise 3.5.2 Show that if $\Omega \subset \mathbb{R}^2$ is a bounded open set with Lipschitz boundary and if $u \in v + W_0^{1,p}(\Omega; \mathbb{R}^2)$, with $p \geq 2$, then

$$\iint_\Omega \det \nabla u \, dx_1 dx_2 = \iint_\Omega \det \nabla v \, dx_1 dx_2.$$

Suggestion: Prove first the result for $u, v \in C^2(\overline{\Omega}; \mathbb{R}^2)$ with $u = v$ on $\partial\Omega$.

Exercise 3.5.3 Let $\Omega \subset \mathbb{R}^2$ be a bounded open set with Lipschitz boundary, $u_0 \in W^{1,p}(\Omega; \mathbb{R}^2)$, with $p \geq 2$, and

$$(P) \quad \inf\left\{ I(u) = \iint_\Omega \det \nabla u(x) \, dx : u \in u_0 + W_0^{1,p}(\Omega; \mathbb{R}^2) \right\} = m.$$

Write the Euler-Lagrange equation associated to (P). Is the result totally surprising?

Exercise 3.5.4 Let $u, v \in W^{1,p}(\Omega; \mathbb{R}^2)$, with $p \geq 2$. Show that there exists $\gamma > 0$ (depending only on p) so that

$$\|\det \nabla u - \det \nabla v\|_{L^{p/2}} \leq \gamma \left(\|\nabla u\|_{L^p} + \|\nabla v\|_{L^p} \right) \|\nabla u - \nabla v\|_{L^p}.$$

Exercise 3.5.5 Let $\Omega \subset \mathbb{R}^2$ be a bounded open set with Lipschitz boundary. We have seen in Lemma 3.23 that, if $p > 2$, then

$$u^\nu \rightharpoonup u \text{ in } W^{1,p}(\Omega; \mathbb{R}^2) \;\Rightarrow\; \det \nabla u^\nu \rightharpoonup \det \nabla u \text{ in } L^{p/2}(\Omega).$$

(i) Show that the result is, in general, false if $p = 2$. To achieve this goal choose, for example, $\Omega = (0,1)^2$ and

$$u^\nu(x_1, x_2) = \frac{1}{\sqrt{\nu}}(1 - x_2)^\nu (\sin \nu x_1, \cos \nu x_1).$$

3.6   Relaxation theory

Recall that the problem under consideration is

$$(P) \quad \inf \left\{ I\left(u \right) = \int_\Omega f\left(x, u\left(x \right), \nabla u\left(x \right) \right) dx : u \in u_0 + W_0^{1,p}\left(\Omega \right) \right\} = m$$

where

- $\Omega \subset \mathbb{R}^n$ is a bounded open set with Lipschitz boundary;
- $f : \overline{\Omega} \times \mathbb{R} \times \mathbb{R}^n \to \mathbb{R}$, $f = f\left(x, u, \xi \right)$, is continuous and non-negative;
- $u_0 \in W^{1,p}\left(\Omega \right)$ with $I\left(u_0 \right) < \infty$.

Before stating the main theorem, let us recall some facts from Section 1.5.

Remark 3.27 The convex envelope of f, with respect to the variable ξ, is denoted by f^{**}. It is the largest convex function (with respect to the variable ξ) which is smaller than f. In other words

$$g\left(x, u, \xi \right) \leq f^{**}\left(x, u, \xi \right) \leq f\left(x, u, \xi \right), \ \forall\left(x, u, \xi \right) \in \overline{\Omega} \times \mathbb{R} \times \mathbb{R}^n$$

for every convex function g (more precisely, $\xi \to g\left(x, u, \xi \right)$ is convex), $g \leq f$. We have two ways of computing this function.

(i) From the duality theorem (Theorem 1.56) we have, for every $\left(x, u, \xi \right) \in \overline{\Omega} \times \mathbb{R} \times \mathbb{R}^n$,

$$f^*\left(x, u, \xi^* \right) = \sup_{\xi \in \mathbb{R}^n} \left\{ \langle \xi; \xi^* \rangle - f\left(x, u, \xi \right) \right\}$$

$$f^{**}\left(x, u, \xi \right) = \sup_{\xi^* \in \mathbb{R}^n} \left\{ \langle \xi; \xi^* \rangle - f^*\left(x, u, \xi^* \right) \right\}.$$

(ii) From Carathéodory theorem (Theorem 1.57) we have, for every $\left(x, u, \xi \right) \in \overline{\Omega} \times \mathbb{R} \times \mathbb{R}^n$,

$$f^{**}\left(x, u, \xi \right) = \inf \left\{ \sum_{i=1}^{n+1} \lambda_i f\left(x, u, \xi_i \right) : \xi = \sum_{i=1}^{n+1} \lambda_i \xi_i, \ \lambda_i \geq 0 \text{ and } \sum_{i=1}^{n+1} \lambda_i = 1 \right\}.$$

(iii) In general, the function $f^{**} : \overline{\Omega} \times \mathbb{R} \times \mathbb{R}^n \to \mathbb{R}$ is, however, not continuous in the variables $\left(x, u \right)$ (see Exercise 3.6.7); of course, since f^{**} is convex in the variable ξ, it is continuous with respect to this variable. If the function f has the following structure

$$f\left(x, u, \xi \right) = f_1\left(x, u \right) f_2\left(\xi \right) + f_3\left(x, u \right)$$

with $f_1, f_3 : \overline{\Omega} \times \mathbb{R} \to \mathbb{R}$ continuous and $f_1 \geq 0$, then

$$f^{**}\left(x, u, \xi \right) = f_1\left(x, u \right) f_2^{**}\left(\xi \right) + f_3\left(x, u \right)$$

We conclude this section with two examples.

Example 3.30 Let us return to Bolza example (Example 3.10). Here we have $n = 1$,

$$f(x, u, \xi) = f(u, \xi) = (\xi^2 - 1)^2 + u^4$$

$$(P) \quad \inf \left\{ I(u) = \int_0^1 f(u(x), u'(x)) \, dx : u \in W_0^{1,4}(0,1) \right\} = m.$$

We have already shown that $m = 0$ and that (P) has no solution. An elementary computation (cf. Example 1.55 (ii)) shows that

$$f^{**}(u, \xi) = \begin{cases} f(u, \xi) & \text{if } |\xi| \geq 1 \\ u^4 & \text{if } |\xi| < 1. \end{cases}$$

Therefore $\overline{u} \equiv 0$ is a solution of

$$(\overline{P}) \quad \inf \left\{ \overline{I}(u) = \int_0^1 f^{**}(u(x), u'(x)) \, dx : u \in W_0^{1,4}(0,1) \right\} = \overline{m} = 0.$$

The sequence $u_\nu \in W_0^{1,4}$ ($\nu \geq 2$ being an integer) constructed in Example 3.10 satisfies the conclusions of the theorem, i.e.

$$u_\nu \rightharpoonup \overline{u} \text{ in } W^{1,4} \quad \text{and} \quad I(u_\nu) \to \overline{I}(\overline{u}) = 0, \text{ as } \nu \to \infty.$$

Example 3.31 Let $\Omega \subset \mathbb{R}^2$ be a bounded open set with Lipschitz boundary. Let $u_0 \in W^{1,4}(\Omega; \mathbb{R}^2)$ be such that

$$\iint_\Omega \det \nabla u_0(x) \, dx \neq 0.$$

Let, for $\xi \in \mathbb{R}^{2 \times 2}$, $f(\xi) = (\det \xi)^2$,

$$(P) \quad \inf \left\{ I(u) = \iint_\Omega f(\nabla u(x)) \, dx : u \in u_0 + W_0^{1,4}(\Omega; \mathbb{R}^2) \right\} = m$$

$$(\overline{P}) \quad \inf \left\{ \overline{I}(u) = \iint_\Omega f^{**}(\nabla u(x)) \, dx : u \in u_0 + W_0^{1,4}(\Omega; \mathbb{R}^2) \right\} = \overline{m}.$$

Let us show that the conclusion of Theorem 3.28 is false in this context, by proving that $m > \overline{m}$. Indeed it is easy to prove (cf. Exercise 1.5.5) that $f^{**}(\xi) \equiv 0$, which therefore implies that $\overline{m} = 0$. Let us show that $m > 0$. Indeed by Jensen inequality (cf. Theorem 1.52) we have, for every $u \in u_0 + W_0^{1,4}(\Omega; \mathbb{R}^2)$,

$$\iint_\Omega (\det \nabla u(x))^2 \, dx \geq \text{meas } \Omega \left(\frac{1}{\text{meas } \Omega} \iint_\Omega \det \nabla u(x) \, dx \right)^2.$$

Exercise 3.6.4 Let $\Omega \subset \mathbb{R}^2$ be a bounded open set with Lipschitz boundary,

$$f(x, u, \xi) = f(\xi) = \left((\xi_1)^2 - 1\right)^2 + (\xi_2)^4$$

where $\xi = (\xi_1, \xi_2) \in \mathbb{R}^2$ and

$$(P) \quad \inf\left\{I(u) = \iint_\Omega f(\nabla u(x_1, x_2)) \, dx_1 dx_2 : u \in W_0^{1,4}(\Omega)\right\} = m.$$

Evaluate m and show, with the help of Exercise 1.4.11, that (P) has no solution.

Exercise 3.6.5 Let

$$\Omega = \left\{(x_1, x_2) \in \mathbb{R}^2 : |x_1 - x_2|, |x_1 + x_2| < 1\right\}$$

$$f(x, u, \xi) = f(\xi) = \left((\xi_1)^2 - 1\right)^2 + \left((\xi_2)^2 - 1\right)^2$$

where $\xi = (\xi_1, \xi_2) \in \mathbb{R}^2$ and

$$(P) \quad \inf\left\{I(u) = \iint_\Omega f(\nabla u(x_1, x_2)) \, dx_1 dx_2 : u \in W_0^{1,4}(\Omega)\right\} = m.$$

Compute m and prove that

$$\overline{u}(x_1, x_2) = 1 - \max\left\{|x_1 - x_2|, |x_1 + x_2|\right\}$$

is a solution of (P).

Exercise 3.6.6 Let $f(\xi) = e^{-|\xi|}$ and

$$(P) \quad \inf\left\{I(u) = \int_0^1 f(u'(x)) \, dx : u \in W_0^{1,1}(0,1)\right\} = m.$$

Prove, with the help of Exercise 3.6.2, that $m = 0$. Show that there is no sequence $u_\nu \in W_0^{1,1}(\Omega)$ so that

$$u_\nu \rightharpoonup u = 0 \text{ in } W^{1,1} \quad \text{and} \quad I(u_\nu) \to m = \overline{I}(0) = 0, \text{ as } \nu \to \infty$$

although there is a sequence satisfying

$$u_\nu \to u = 0 \text{ in } L^\infty \quad \text{and} \quad I(u_\nu) \to m = \overline{I}(0) = 0, \text{ as } \nu \to \infty.$$

The Euler-Lagrange equation is then $\Delta u = -h$. When $n = 1$, the assumption $h \in C^{k,\alpha}$ or $W^{k,p}$ with $k \geq 0$ and $0 \leq \alpha \leq 1$ or $1 \leq p \leq \infty$, immediately implies $u \in C^{k+2,\alpha}$ or $W^{k+2,p}$. When $n \geq 2$, the equation $\Delta u = -h$ states that the trace of the Hessian matrix is in $C^{k,\alpha}$ or $W^{k,p}$ and the regularity that we want to establish is that *each* entry of the Hessian matrix is in $C^{k,\alpha}$ or $W^{k,p}$. Presented like this, the result seems unlikely. It is one of the achievements of regularity theory to prove that this is indeed the case, provided $0 < \alpha < 1$ or $1 < p < \infty$; the limit cases $\alpha = 0, 1$ or $p = 1, \infty$ turning out to be false as seen in the exercises.

We now detail the contents of Sections 4.3 to 4.7. We proceed in several steps and restrict our attention to the case $p = 2$. The first step is to obtain one degree smoother solutions, namely to show that $\overline{u} \in W^{2,2}$. This is achieved in Section 4.3, where we obtain *interior regularity* and in Section 4.4, under some smoothness of the boundary of the domain Ω, we get *regularity up to the boundary*. The main tool in both cases is the so-called *difference quotient method*.

To get even smoother solutions, we have to restrict our attention to the Dirichlet integral or more generally to quadratic integrands. This is dealt with in Section 4.5. In Section 4.6, we give a completely different proof of the regularity of solutions in the case of Dirichlet integral.

In Section 4.7 we provide, without proofs, some general theorems on higher regularity.

We should also point out that all the regularity results that we obtain here are about solutions of the Euler-Lagrange equation and therefore not only minimizers of (P).

The problem of regularity, including the closely related ones concerning regularity for elliptic partial differential equations, is a difficult one that has attracted many mathematicians. We quote only a few of them: Agmon, Bernstein, Calderon, De Giorgi, Douglis, E. Hopf, Leray, Lichtenstein, Morrey, Moser, Nash, Nirenberg, Rado, Schauder, Tonelli, Weyl and Zygmund.

In addition to the books that were mentioned in Chapter 3 one can consult those by Evans [44], Gilbarg-Trudinger [51] and Ladyzhenskaya-Uraltseva [70] (for Dirichlet integral see also, for example, Brézis [14], Folland [47] and John [67]).

4.2   The one dimensional case

Let us restate the problem. We consider

$$(P) \quad \inf_{u \in X} \left\{ I(u) = \int_a^b f(x, u(x), u'(x))\, dx \right\} = m$$

To prove further regularity of $\overline{u}$, we start by showing that $\overline{u} \in W^{2,2}(a,b)$. This follows immediately from (4.1) and from the definition of weak derivative. Indeed since $\overline{u} \in W^{1,2}$, we have that $\overline{u} \in L^\infty$ and thus $g_u(x, \overline{u}) \in L^2$, leading to

$$\left| \int_a^b \overline{u}'v' \, dx \right| \leq \|g_u(x,\overline{u})\|_{L^2} \|v\|_{L^2} , \ \forall v \in C_0^\infty(a,b). \tag{4.2}$$

Theorem 1.37 implies then that $\overline{u} \in W^{2,2}$. We can then integrate by parts (4.1), bearing in mind that $v(a) = v(b) = 0$, and using the fundamental lemma of the calculus of variations (cf. Theorem 1.24), we deduce that

$$\overline{u}''(x) = g_u(x, \overline{u}(x)), \ \text{a.e. } x \in (a,b). \tag{4.3}$$

We are now in a position to start an iteration process. Since $\overline{u} \in W^{2,2}(a,b)$ we deduce that (cf. Theorem 1.43) $\overline{u} \in C^1([a,b])$ and hence the function

$$x \to g_u(x, \overline{u}(x))$$

is $C^1([a,b])$, g being C^∞. Returning to (4.3) we deduce that $\overline{u}'' \in C^1$ and hence $\overline{u} \in C^3$.

From there we can infer that $x \to g_u(x, \overline{u}(x))$ is C^3, and thus from (4.3) we obtain that $\overline{u}'' \in C^3$ and hence $\overline{u} \in C^5$.

Continuing this process we have indeed established that $\overline{u} \in C^\infty([a,b])$. ∎

We now generalize the argument of the proposition and we start with a lemma.

Lemma 4.2 *Let $f \in C^1([a,b] \times \mathbb{R} \times \mathbb{R})$ satisfy (H_1), (H_2) and (H_3'). Then any minimizer $\overline{u} \in W^{1,p}(a,b)$ of (P) is in fact in $W^{1,\infty}(a,b)$ and the Euler-Lagrange equation holds almost everywhere, i.e.*

$$\frac{d}{dx}[f_\xi(x,\overline{u},\overline{u}')] = f_u(x,\overline{u},\overline{u}'), \ \text{a.e. } x \in (a,b).$$

Proof. We know from Theorem 3.11 and Remark 3.12 that the following equation holds

$$(E_w) \quad \int_a^b [f_u(x,\overline{u},\overline{u}')v + f_\xi(x,\overline{u},\overline{u}')v'] \, dx = 0, \ \forall v \in C_0^\infty(a,b). \tag{4.4}$$

We then divide the proof into two steps.

Step 1. Define

$$\varphi(x) = f_\xi(x,\overline{u}(x),\overline{u}'(x)) \quad \text{and} \quad \psi(x) = f_u(x,\overline{u}(x),\overline{u}'(x)).$$

Proof. We propose a different proof in Exercise 4.2.1. The present one is more direct and uses Lemma 2.9.

Step 1. We know from Lemma 4.2 that $x \to \varphi(x) = f_\xi(x, \overline{u}(x), \overline{u}'(x))$ is in $W^{1,1}(a, b)$ and hence it is continuous. Appealing to Lemma 2.9 (and the remark following this lemma), we have that if

$$H(x, u, v) = \sup_{\xi \in \mathbb{R}} \{v\,\xi - f(x, u, \xi)\}$$

then $H \in C^\infty([a, b] \times \mathbb{R} \times \mathbb{R})$ and, for every $x \in [a, b]$, we have

$$\varphi(x) = f_\xi(x, \overline{u}(x), \overline{u}'(x)) \quad \Leftrightarrow \quad \overline{u}'(x) = H_v(x, \overline{u}(x), \varphi(x)).$$

Since H_v, $\overline{u}$ and φ are continuous, we infer that $\overline{u}'$ is continuous and hence $\overline{u} \in C^1([a, b])$. We therefore deduce that $x \to f_u(x, \overline{u}(x), \overline{u}'(x))$ is continuous, which combined with the fact that (cf. (4.5))

$$\frac{d}{dx}[\varphi(x)] = f_u(x, \overline{u}(x), \overline{u}'(x)), \text{ a.e. } x \in (a, b)$$

(or equivalently, by Lemma 2.9, $\varphi' = -H_u(x, \overline{u}, \varphi)$) leads to $\varphi \in C^1([a, b])$.

Step 2. Returning to our Hamiltonian system

$$\begin{cases} \overline{u}'(x) = H_v(x, \overline{u}(x), \varphi(x)) \\ \varphi'(x) = -H_u(x, \overline{u}(x), \varphi(x)) \end{cases}$$

we can start our iteration. Indeed since H is C^∞ and $\overline{u}$ and φ are C^1 we deduce from our system that, in fact, $\overline{u}$ and φ are C^2. Returning to the system we get that $\overline{u}$ and φ are C^3. Finally we get that $\overline{u}$ is C^∞, as wished. ∎

We conclude the section by giving an example where we can get further regularity without assuming the non-degeneracy condition $f_{\xi\xi} > 0$.

Theorem 4.5 *Let $g \in C^1([a, b] \times \mathbb{R})$ satisfy*

(H_2) *there exist $p > q \geq 1$ and $\alpha_2, \alpha_3 \in \mathbb{R}$ such that*

$$g(x, u) \geq \alpha_2|u|^q + \alpha_3, \ \forall(x, u) \in [a, b] \times \mathbb{R}.$$

Let

$$f(x, u, \xi) = \frac{1}{p}|\xi|^p + g(x, u).$$

Then there exists $\overline{u} \in C^1([a, b])$, with $|\overline{u}'|^{p-2}\,\overline{u}' \in C^1([a, b])$, a minimizer of (P) and the Euler-Lagrange equation holds everywhere, i.e.

$$\frac{d}{dx}\left[|\overline{u}'(x)|^{p-2}\,\overline{u}'(x)\right] = g_u(x, \overline{u}(x)), \ \forall x \in [a, b].$$

Exercise 4.2.3 Let $p > 2q > 2$ and

$$f(x, u, \xi) = f(u, \xi) = \frac{1}{p} |\xi|^p + \frac{\lambda}{q} |u|^q \text{ where } \lambda = \frac{q p^{q-1} (p-1)}{(p-q)^q}$$

$$\overline{u}(x) = \frac{p-q}{p} |x|^{p/(p-q)}$$

(note that if, for example, $p = 6$ and $q = 2$, then $f \in C^\infty (\mathbb{R}^2)$).

   (i) Show that $\overline{u} \in C^1 ([-1, 1])$ but $\overline{u} \notin C^2 ([-1, 1])$.

   (ii) Find some values of p and q so that

$$|\overline{u}'|^{p-2} \overline{u}', |\overline{u}|^{q-2} \overline{u} \in C^\infty ([-1, 1]),$$

although $\overline{u} \notin C^2 ([-1, 1])$.

   (iii) Show that $\overline{u}$ is the unique minimizer of

$$(P) \quad \inf_{u \in W^{1,p}(-1,1)} \left\{ I(u) = \int_{-1}^1 f(u(x), u'(x)) \, dx : u(-1) = u(1) = \frac{p-q}{p} \right\}.$$

Exercise 4.2.4 Let

$$\varphi(x) = \begin{cases} \exp(-1/x^2) & \text{if } x \neq 0 \\ 0 & \text{if } x = 0 \end{cases}$$

$$f(x, u, \xi) = f(x, \xi) = \left[\varphi(x) \xi - 2x \varphi(x) \sin \frac{\pi}{x} + \pi \varphi(x) \cos \frac{\pi}{x} \right]^2$$

and

$$(P) \quad \inf \left\{ I(u) = \int_0^1 f(x, u'(x)) \, dx : u \in W_0^{1,2}(-1, 1) \right\} = m.$$

Observe that $f \in C^\infty ([-1, 1] \times \mathbb{R})$ and $\xi \to f(x, \xi)$ is convex with $f_{\xi\xi}(x, \xi) = 2\varphi(x) > 0$ except at $x = 0$. Show that

$$\overline{u}(x) = \begin{cases} x^2 \sin(\pi/x) & \text{if } x \neq 0 \\ 0 & \text{if } x = 0 \end{cases}$$

is the unique minimizer of (P), $\overline{u} \in W_0^{1,\infty}(-1, 1)$ but $\overline{u} \notin C^1 ([-1, 1])$.

$$(G_4) \quad \left| \frac{\partial^2 g}{\partial x_i \partial \xi_j} (x, \xi) \right| \le \alpha_6 |\xi|, \; \forall (x, \xi) \in \overline{\Omega} \times \mathbb{R}^n \quad and \quad \forall i, j = 1, \cdots, n$$

then the minimizer $u \in u_0 + W_0^{1,2}(\Omega)$ is unique and $u \in W_{loc}^{2,2}(\Omega)$. More precisely for every open set $O \subset \overline{O} \subset \Omega$ there exists a constant $\gamma > 0$ such that

$$\|u\|_{W^{2,2}(O)} \le \gamma \left[\|h\|_{L^2(\Omega)} + \|u\|_{W^{1,2}(\Omega)} \right]. \tag{4.8}$$

Moreover the following form of the Euler-Lagrange equation is satisfied, namely

$$\sum_{i=1}^{n} \frac{\partial}{\partial x_i} \left[\frac{\partial g}{\partial \xi_i} (x, \nabla u) \right] = -h, \; a.e. \; in \; \Omega. \tag{4.9}$$

Proof. The existence of a minimizer follows at once from Theorem 3.3 (and Remark 3.4 (i)). The Euler-Lagrange equation is established via Theorem 3.11 and Remark 3.12 (ii), once it has been observed that (G_1) implies (cf. Exercise 1.5.8) the existence of a constant $\alpha_7 > 0$ such that

$$\left| \frac{\partial g}{\partial \xi_i} (x, \xi) \right| \le \alpha_7 (|\xi| + 1), \; \forall (x, \xi) \in \overline{\Omega} \times \mathbb{R}^n \text{ and } \forall i = 1, \cdots, n. \tag{4.10}$$

The uniqueness in Part 2 is a consequence of (G_3) which implies the strict convexity of g (see Exercise 3.3.4). The equation (4.9) is deduced from (4.7) by integration by parts, once it has been established that $u \in W_{loc}^{2,2}(\Omega)$. It therefore remains to prove this last fact and the appropriate estimate. We divide the proof into two steps.

Step 1. We here recall three elementary facts that are constantly used in the proof of the theorem.

Fact 1. We recall (cf. Notation 1.36) the notion of *difference quotient*. For $\tau \in \mathbb{R}^n$, $\tau \neq 0$, we let

$$(D_\tau u)(x) = \frac{u(x + \tau) - u(x)}{|\tau|}.$$

It is clear that

$$\nabla (D_\tau u) = D_\tau (\nabla u).$$

Recall next that Theorem 1.37 asserts the following properties for every open set $\Omega \subset \mathbb{R}^n$.

(i) For every open set $\omega \subset \overline{\omega} \subset \Omega$, with $\overline{\omega}$ compact and for every $u \in W^{1,2}(\Omega)$, the following holds

$$\|D_\tau u\|_{L^2(\omega)} \le \|\nabla u\|_{L^2(\Omega)} \tag{4.11}$$

We then make a special choice of $\varphi \in W_0^{1,2}(\Omega)$ in the Euler-Lagrange equation (4.7), namely, for $|\tau|$ sufficiently small,

$$\varphi = D_{-\tau}\left(\rho^2 D_\tau u\right).$$

Equation (4.7) becomes then

$$\int_\Omega \langle \nabla_\xi g\left(x, \nabla u\right) ; \nabla \left(D_{-\tau}\rho^2 \left(D_\tau u\right)\right)\rangle = \int_\Omega h D_{-\tau}\left(\rho^2 D_\tau u\right).$$

Invoking the properties of the difference quotient, notably (4.13), we deduce that

$$\int_\Omega \left\langle D_\tau \left[\nabla_\xi g\left(x, \nabla u\right)\right] ; \rho^2 \left(D_\tau \nabla u\right)\right\rangle + 2\int_\Omega \left\langle D_\tau \left[\nabla_\xi g\left(x, \nabla u\right)\right] ; \rho \nabla \rho \left(D_\tau u\right)\right\rangle$$

$$= \int_\Omega h D_{-\tau}\left(\rho^2 D_\tau u\right) \tag{4.15}$$

where

$$D_\tau \left[\nabla_\xi g\left(x, \nabla u\right)\right] = \left(D_\tau \left[\frac{\partial g}{\partial \xi_1}\left(x, \nabla u\right)\right], \cdots, D_\tau \left[\frac{\partial g}{\partial \xi_n}\left(x, \nabla u\right)\right]\right)$$

and

$$D_\tau \left[\frac{\partial g}{\partial \xi_i}\left(x, \nabla u\right)\right] = \frac{1}{|\tau|}\left[\frac{\partial g}{\partial \xi_i}\left(x + \tau, \nabla u\left(x + \tau\right)\right) - \frac{\partial g}{\partial \xi_i}\left(x, \nabla u\left(x\right)\right)\right].$$

We then estimate each one of the three terms separately and this will lead us to the desired estimate in Step 2.2; but we start with a preliminary computation.

Step 2.1. We point out two simple observations.

Observation 1. Note that

$$\frac{\partial g}{\partial \xi_i}\left(x + \tau, \nabla u\left(x + \tau\right)\right) - \frac{\partial g}{\partial \xi_i}\left(x + \tau, \nabla u\left(x\right)\right)$$

$$= \int_0^1 \frac{d}{dt}\left[\frac{\partial g}{\partial \xi_i}\left(x + \tau, \nabla u\left(x\right) + t\left(\nabla u\left(x + \tau\right) - \nabla u\left(x\right)\right)\right)\right] dt$$

$$= \sum_{j=1}^n \int_0^1 \left[\frac{\partial^2 g}{\partial \xi_i \partial \xi_j}\left(x + \tau, \nabla u\left(x\right) + t\left(\nabla u\left(x + \tau\right) - \nabla u\left(x\right)\right)\right)\right.$$

$$\left.\left(u_{x_j}\left(x + \tau\right) - u_{x_j}\left(x\right)\right)\right] dt.$$

Observation 2. We next see that

$$\frac{\partial g}{\partial \xi_i}\left(x + \tau, \nabla u\left(x\right)\right) - \frac{\partial g}{\partial \xi_i}\left(x, \nabla u\left(x\right)\right) = \int_0^1 \frac{d}{dt}\left[\frac{\partial g}{\partial \xi_i}\left(x + t\tau, \nabla u\left(x\right)\right)\right] dt$$

$$= \int_0^1 \sum_{j=1}^n \frac{\partial^2 g}{\partial \xi_i \partial x_j}\left(x + t\tau, \nabla u\left(x\right)\right) \tau_j \, dt.$$

Estimate 2. From (4.16), we have, for $\epsilon > 0$ fixed,

$$\left| 2 \int_\Omega \langle D_\tau \left[\nabla_\xi g \left(x, \nabla u \right) \right] ; \rho \nabla \rho \left(D_\tau u \right) \rangle \right|$$

$$\leq 2n\alpha_4 \int_\Omega \left| \rho \left(D_\tau \nabla u \right) \right| \left| \nabla \rho \left(D_\tau u \right) \right| + 2n\alpha_6 \int_\Omega \left| \rho \nabla u \right| \left| \nabla \rho \left(D_\tau u \right) \right|$$

$$\leq 2n\alpha_4 \epsilon \left\| \rho \left(D_\tau \nabla u \right) \right\|_{L^2}^2 + \left(\frac{2n\alpha_4}{\epsilon} + 2n\alpha_6 \right) \left\| \nabla \rho \left(D_\tau u \right) \right\|_{L^2}^2 + 2n\alpha_6 \left\| \rho \nabla u \right\|_{L^2}^2 .$$

Observing that, from (4.11), we have

$$\left\| \nabla \rho \left(D_\tau u \right) \right\|_{L^2}^2 = \int_{\text{supp } \rho} \left| \nabla \rho \left(D_\tau u \right) \right|^2 \leq \left\| \nabla \rho \right\|_{L^\infty}^2 \int_{\text{supp } \rho} \left| D_\tau u \right|^2 \leq \left\| \nabla \rho \right\|_{L^\infty}^2 \left\| \nabla u \right\|_{L^2}^2 \tag{4.18}$$

and we thus get, recalling that $0 \leq \rho \leq 1$,

$$\left| 2 \int_\Omega \langle D_\tau \left[\nabla_\xi g \left(x, \nabla u \right) \right] ; \rho \nabla \rho \left(D_\tau u \right) \rangle \right|$$

$$\leq 2n\alpha_4 \epsilon \left\| \rho \left(D_\tau \nabla u \right) \right\|_{L^2}^2 + \left(\frac{2n\alpha_4}{\epsilon} + 2n\alpha_6 \right) \left\| \nabla \rho \right\|_{L^\infty}^2 \left\| \nabla u \right\|_{L^2}^2 + 2n\alpha_6 \left\| \nabla u \right\|_{L^2}^2 .$$

Estimate 3. It follows from (4.11) that, for $\epsilon > 0$ fixed,

$$\left| \int_\Omega h \left(D_{-\tau} \rho^2 \left(D_\tau u \right) \right) \right| \leq \epsilon \left\| \left(D_{-\tau} \rho^2 \left(D_\tau u \right) \right) \right\|_{L^2}^2 + \frac{1}{\epsilon} \left\| h \right\|_{L^2}^2$$

$$\leq \epsilon \left\| \nabla \left(\rho^2 \left(D_\tau u \right) \right) \right\|_{L^2}^2 + \frac{1}{\epsilon} \left\| h \right\|_{L^2}^2 .$$

Since

$$\left\| \nabla \left(\rho^2 \left(D_\tau u \right) \right) \right\|_{L^2}^2 = \int_\Omega \left| \rho^2 \left(D_\tau \nabla u \right) + 2\rho \nabla \rho \left(D_\tau u \right) \right|^2$$

$$\leq 2 \int_\Omega \left| \rho^2 \left(D_\tau \nabla u \right) \right|^2 + 2 \int_\Omega \left| 2\rho \nabla \rho \left(D_\tau u \right) \right|^2$$

we obtain, as in (4.18) and recalling that $0 \leq \rho \leq 1$ and thus $\rho^2 \leq \rho$,

$$\left| \int_\Omega h \left(D_{-\tau} \rho^2 \left(D_\tau u \right) \right) \right| \leq 2\epsilon \left\| \rho \left(D_\tau \nabla u \right) \right\|_{L^2}^2 + 8\epsilon \left\| \nabla \rho \right\|_{L^\infty}^2 \left\| \nabla u \right\|_{L^2}^2 + \frac{1}{\epsilon} \left\| h \right\|_{L^2}^2 .$$

4.4   The difference quotient method: boundary regularity

We now improve Theorem 4.7 in a special case, so as to get regularity up to the boundary, i.e. $u \in W^{2,2}$ and not only in $W^{2,2}_{\text{loc}}$.

Theorem 4.8 *Let $\Omega \subset \mathbb{R}^n$ be a bounded open set with C^2 boundary, $h \in L^2 (\Omega)$ and $g_{ij} = g_{ji} \in C^1 (\overline{\Omega})$ such that there exists $\alpha > 0$ with*

$$(G) \quad \sum_{i,j=1}^{n} g_{ij} (x) \lambda_i \lambda_j \geq \alpha |\lambda|^2 , \; \forall \lambda \in \mathbb{R}^n \;\; and \;\; \forall x \in \overline{\Omega}.$$

Then there exists a unique minimizer $u \in W^{1,2}_0 (\Omega) \cap W^{2,2} (\Omega)$ of

$$(P) \quad \inf \left\{ I (u) = \sum_{i,j=1}^{n} \int_\Omega g_{ij} u_{x_i} u_{x_j} dx - \int_\Omega h\, u\, dx : u \in W^{1,2}_0 (\Omega) \right\}$$

and there exists a constant $\gamma > 0$ such that

$$\|u\|_{W^{2,2}(\Omega)} \leq \gamma \|h\|_{L^2(\Omega)} . \tag{4.19}$$

Remark 4.9 The theorem can be extended to the non-quadratic setting of Theorem 4.7.

Proof. The proof is in the same spirit as that of Theorem 4.7 but technically heavier. We only outline the main steps. We refer for more details to Brézis [14], John [67] or Evans [44] that we follow here. The Euler-Lagrange equation is then

$$2 \int_\Omega \sum_{i,j=1}^{n} g_{ij} u_{x_i} \varphi_{x_j} = \int_\Omega h\varphi, \; \forall \varphi \in W^{1,2}_0 (\Omega) \tag{4.20}$$

and, since $u \in W^{2,2}_{\text{loc}} (\Omega)$, we also have, after integration by parts and using Theorem 1.24,

$$\sum_{i,j=1}^{n} g_{ij} u_{x_i x_j} + \sum_{i,j=1}^{n} \frac{\partial}{\partial x_j} [g_{ij}] u_{x_i} = -\frac{h}{2}, \; \text{a.e. in } \Omega. \tag{4.21}$$

Step 1. We start with a special choice of Ω, namely

$$\Omega = Q_+ = Q \cap \{x_n > 0\}$$

So let $W \subset \mathbb{R}^n$ be a bounded open set such that there exists a one-to-one and onto map $H : \overline{Q} \to \overline{W}$ such that

$$H \in C^2\left(\overline{Q}\right), \quad H^{-1} \in C^2\left(\overline{W}\right) \quad \text{and} \quad \det \nabla H > 0.$$

Let $W_+ = H\left(Q_+\right)$, $w \subset \overline{w} \subset W$ be open and $w_+ = w \cap W_+$. We want to prove that we can find $\gamma_4 > 0$ such that

$$\|u\|^2_{W^{2,2}(w_+)} \le \gamma_4 \left(\|u\|^2_{W^{1,2}(W_+)} + \|h\|^2_{L^2(W_+)}\right). \tag{4.24}$$

We change variables and set

$$x = H\left(y\right), \ u\left(x\right) = v\left(H^{-1}\left(x\right)\right), \ \varphi\left(x\right) = \psi\left(H^{-1}\left(x\right)\right).$$

We therefore immediately have (see Exercise 4.4.1), with the change of variables $x = H\left(y\right)$,

$$\sum_{i,j=1}^{n} \int_{W_+} g_{ij}\left(x\right) u_{x_i}\left(x\right) \varphi_{x_j}\left(x\right) dx = \sum_{k,l=1}^{n} \int_{Q_+} \widetilde{g}_{kl}\left(y\right) v_{y_k}\left(y\right) \psi_{y_l}\left(y\right) dy$$

where

$$\widetilde{g}_{kl}\left(y\right) = \sum_{i,j=1}^{n} g_{ij}\left(H\left(y\right)\right) \frac{\partial H_k^{-1}}{\partial x_i}\left(H\left(y\right)\right) \frac{\partial H_l^{-1}}{\partial x_j}\left(H\left(y\right)\right) \det \nabla H\left(y\right).$$

Similarly, setting $\widetilde{h}\left(y\right) = h\left(H\left(y\right)\right) \det \nabla H\left(y\right)$, we infer that

$$\int_{W_+} h\left(x\right) \varphi\left(x\right) dx = \int_{Q_+} \widetilde{h}\left(y\right) \psi\left(y\right) dy.$$

The Euler-Lagrange equation (4.20) then becomes

$$2 \sum_{k,l=1}^{n} \int_{Q_+} \widetilde{g}_{kl}\left(y\right) v_{y_k}\left(y\right) \psi_{y_l}\left(y\right) dy = \int_{Q_+} \widetilde{h}\left(y\right) \psi\left(y\right) dy, \ \forall \psi \in W_0^{1,2}\left(Q_+\right).$$
$$\tag{4.25}$$

Note (see Exercise 4.4.1) that the uniform ellipticity is preserved, which means that there exists a constant $\beta > 0$ such that

$$\sum_{k,l=1}^{n} \widetilde{g}_{kl}\left(y\right) \lambda_k \lambda_l \ge \beta \left|\lambda\right|^2, \ \forall \lambda \in \mathbb{R}^n \text{ and } \forall y \in \overline{Q}_+.$$

We then apply Step 1 to (4.25) to get (4.23), namely

$$\|v\|^2_{W^{2,2}(q_+)} \le \widetilde{\gamma}_3 \left(\|v\|^2_{W^{1,2}(Q_+)} + \left\|\widetilde{h}\right\|^2_{L^2(Q_+)}\right)$$

leads to

$$\sum_{i,j=1}^{n} \int_{U} a_{ij}(x) u_{x_i}(x) \varphi_{x_j}(x) \, dx = \sum_{k,l=1}^{n} \int_{Q} b_{kl}(y) v_{y_k}(y) \psi_{y_l}(y) \, dy$$

where

$$b_{kl}(y) = \sum_{i,j=1}^{n} a_{ij}(H(y)) \frac{\partial H_k^{-1}}{\partial x_i}(H(y)) \frac{\partial H_l^{-1}}{\partial x_j}(H(y)) \det \nabla H(y).$$

Moreover the following holds

$$\sum_{k,l=1}^{n} b_{kl}(y) \lambda_k \lambda_l \geq \beta |\lambda|^2, \ \forall \lambda \in \mathbb{R}^n \text{ and } \forall y \in \overline{Q}$$

for a certain constant $\beta > 0$.

4.5   Higher regularity for the Dirichlet integral

To get higher regularity in the context of Sections 4.3 and 4.4 is a difficult task and requires new ideas which are explained in Section 4.7 (cf. Theorem 4.16). However when we are considering Dirichlet integral or more generally quadratic integrands with regular coefficients (cf. Exercise 4.5.2) this is much easier and we discuss this matter now. Let us first express informally the procedure. Once we have established the fact that $u \in W^{2,2}$ and satisfies the equation

$$\Delta u = -h$$

it is enough to differentiate with respect to any of the variables to have the equation

$$\Delta u_{x_i} = -h_{x_i}$$

and restart the process to get $u_{x_i} \in W^{2,2}$ and thus $u \in W^{3,2}$. Iterating the process we have the maximal possible regularity.

Let us now be more precise. We recall that $\Omega \subset \mathbb{R}^n$ is a bounded open set with Lipschitz boundary, $h \in L^2(\Omega)$ and

$$(P) \quad \inf \left\{ I(u) = \frac{1}{2} \int_{\Omega} |\nabla u(x)|^2 \, dx - \int_{\Omega} h(x) u(x) \, dx : u \in W_0^{1,2}(\Omega) \right\}.$$

We have seen in Theorem 4.7 that there exists a unique minimizer $u \in W_0^{1,2}(\Omega) \cap W_{\text{loc}}^{2,2}(\Omega)$ of (P). We now have the following result.

$$\|u\|_{W^{2,2}(O)} \leq \gamma_1 \|h\|_{L^2(\Omega)} \quad \text{and} \quad \Delta u = -h, \text{ a.e. in } \Omega. \tag{4.30}$$

Let now $h \in W^{1,2}(\Omega)$ and let us show that $u \in W^{3,2}_{\text{loc}}(\Omega)$. The general case $h \in W^{k,2}$ implying that $u \in W^{k+2,2}_{\text{loc}}(\Omega)$ follows by repeating the argument. The idea is simple, it consists in applying Theorem 4.7 (more precisely the regularity part for the solution of the Euler-Lagrange equation (4.7)) to $u_{x_i} = \partial u/\partial x_i$ and observing that since $\Delta u = -h$, then $\Delta u_{x_i} = -h_{x_i}$. Indeed it is elementary to see that we have, for every $i = 1, \cdots, n$,

$$\int_\Omega \langle \nabla u_{x_i}(x); \nabla\varphi(x) \rangle \, dx = \int_\Omega h_{x_i}(x) \varphi(x) \, dx, \ \forall \varphi \in W^{1,2}_0(\Omega).$$

To prove this, it is sufficient to establish it for $\varphi \in C^\infty_0(\Omega)$ (since $C^\infty_0(\Omega)$ is dense in $W^{1,2}_0(\Omega)$). We have, using (4.29), that

$$\int_\Omega \langle \nabla u_{x_i}; \nabla\varphi \rangle \, dx = \int_\Omega \langle (\nabla u)_{x_i}; \nabla\varphi \rangle \, dx = -\int_\Omega \langle \nabla u; (\nabla\varphi)_{x_i} \rangle \, dx$$

$$= -\int_\Omega \langle \nabla u; \nabla\varphi_{x_i} \rangle \, dx = -\int_\Omega h\varphi_{x_i} \, dx = \int_\Omega h_{x_i}\varphi \, dx.$$

Since $h \in W^{1,2}$, we have that $h_{x_i} \in L^2$ and hence, by Theorem 4.7 (and Exercise 4.3.2 or Step 4 of Theorem 4.8), we get that $u_{x_i} \in W^{2,2}_{\text{loc}}$ and there exists a constant $\gamma_2 > 0$ such that

$$\|u_{x_i}\|_{W^{2,2}(O)} \leq \gamma_2 \|h_{x_i}\|_{L^2(\Omega)} \leq \gamma_2 \|h\|_{W^{1,2}(\Omega)}. \tag{4.31}$$

Since this holds for every $i = 1, \cdots, n$, we have indeed obtained that $u \in W^{3,2}_{\text{loc}}$. Moreover the estimate

$$\|u\|_{W^{3,2}(O)} \leq \gamma_3 \|h\|_{W^{1,2}(\Omega)}$$

follows from (4.30) and (4.31). This concludes the proof of the theorem. ∎

4.5.1   Exercises

Exercise 4.5.1 Let $u_0 \in W^{k+2,2}(\Omega)$. Prove that Theorem 4.10 holds for

$$(P) \quad \inf \left\{ I(u) = \frac{1}{2} \int_\Omega |\nabla u(x)|^2 \, dx - \int_\Omega h(x) u(x) \, dx : u \in u_0 + W^{1,2}_0(\Omega) \right\}$$

Exercise 4.5.2 Let $h \in W^{k,2}(\Omega)$, $g_{ij} = g_{ji} \in C^\infty(\overline{\Omega})$ with

$$\sum_{i,j=1}^n g_{ij}(x) \lambda_i \lambda_j \geq \alpha_5 |\lambda|^2, \ \forall x \in \overline{\Omega} \text{ and } \forall \lambda \in \mathbb{R}^n.$$

Theorem 4.12 (Weyl lemma) *Let* $\Omega \subset \mathbb{R}^n$ *be open and* $u \in L^1_{loc}(\Omega)$ *satisfy*

$$\int_\Omega u(x) \, \Delta\varphi(x) \, dx = 0, \ \forall \varphi \in C^\infty_0(\Omega) \tag{4.32}$$

then $u \in C^\infty(\Omega)$ *and* $\Delta u = 0$ *in* Ω.

Remark 4.13 (i) The function u being defined only almost everywhere, we have to interpret the result, as usual, up to a change of the function on a set of measure zero.

(ii) Note that a solution of the weak form of Laplace equation

$$(E_w) \quad \int_\Omega \langle \nabla u(x) ; \nabla\varphi(x) \rangle \, dx = 0, \ \forall \varphi \in W^{1,2}_0(\Omega)$$

satisfies (4.32). The converse being true if, in addition, $u \in W^{1,2}(\Omega)$. Therefore (4.32) can be seen as a "very weak" form of Laplace equation and a solution of this equation as a "very weak" solution of $\Delta u = 0$.

(iii) The case $n = 1$ is elementary and is discussed in Exercise 1.3.8; therefore, from now on, we assume that $n \geq 2$.

Proof. The idea of the proof is to show that, up to redefining the function on a set of measure zero, u is continuous and satisfies the mean value formula

$$u(x) = \frac{n}{\sigma_{n-1} r^n} \int_{B_r(x)} u(y) \, dy \tag{4.33}$$

for every $x \in \Omega$ and $r > 0$ sufficiently small so that

$$B_r(x) = \{y \in \mathbb{R}^n : |y - x| < r\} \subset \overline{B_r(x)} \subset \Omega$$

and where $\sigma_{n-1} = \text{meas}(\partial B_1(0))$ (i.e. $\sigma_1 = 2\pi$, $\sigma_2 = 4\pi, \cdots$). A classical result (cf. Exercise 4.6.1) allows to conclude that in fact $u \in C^\infty(\Omega)$ and hence, by (4.32) and the fundamental lemma of the calculus of variations (cf. Theorem 1.24), $\Delta u = 0$ in Ω, as claimed.

The proof is divided into three steps.

Step 1. In the first step, we prove that a regularization of u satisfies the mean value formula. To achieve this, we first choose an even function $\varphi \in C^\infty_0(B_1(0))$ with $\int \varphi = 1$ and define

$$\varphi_\epsilon(x) = \frac{1}{\epsilon^n} \varphi\left(\frac{x}{\epsilon}\right) \quad \text{and} \quad \Omega_\epsilon = \left\{x \in \Omega : \overline{B_\epsilon(x)} \subset \Omega\right\}.$$

for every $x \in \Omega$ and $r > 0$ sufficiently small so that

$$\overline{B_r(x)} \subset \Omega.$$

Step 3. In order to conclude the proof of the theorem, from Exercise 4.6.1 and (4.32), we only need to check that u is continuous. This is easily seen as follows. Let $x, y \in \Omega$ and r be sufficiently small so that $\overline{B_r(x)} \cup \overline{B_r(y)} \subset \Omega$. We then have that

$$|u(x) - u(y)| = \frac{n}{\sigma_{n-1} r^n} \left| \int_{B_r(x)} u(z)\, dz - \int_{B_r(y)} u(z)\, dz \right|$$

$$\leq \frac{n}{\sigma_{n-1} r^n} \int_O |u(z)|\, dz,$$

where $O = (B_r(x) \cup B_r(y)) \smallsetminus (B_r(x) \cap B_r(y))$. Appealing to the fact that $u \in L^1(B_r(x) \cup B_r(y))$ and to Exercise 1.3.9, we deduce that u is indeed continuous. ∎

4.6.1   Exercise

Exercise 4.6.1 Let $\Omega \subset \mathbb{R}^n$ be an open set and let $\sigma_{n-1} = \text{meas}(\partial B_1(0))$ (i.e. $\sigma_1 = 2\pi$, $\sigma_2 = 4\pi$, $\cdots$). Let $u \in C^0(\Omega)$ satisfy the mean value formula, which states that

$$u(x) = \frac{n}{\sigma_{n-1} r^n} \int_{B_r(x)} u(y)\, dy$$

for every $x \in \Omega$ and for every $r > 0$ sufficiently small so that

$$B_r(x) = \{y \in \mathbb{R}^n : |y - x| < r\} \subset \overline{B_r(x)} \subset \Omega.$$

Show that $u \in C^\infty(\Omega)$.

4.7   Some general results

The generalization of Section 4.5 to integrands of the form $f = f(x, u, \nabla u)$ is a difficult task. We give here, without proof, a general theorem and we refer for more results to the literature. The next theorem can be found in Morrey [79] (Theorem 1.10.4).

Theorem 4.14 *Let $\Omega \subset \mathbb{R}^n$ be a bounded open set and $f \in C^\infty(\Omega \times \mathbb{R} \times \mathbb{R}^n)$, $f = f(x, u, \xi)$. Let $f_x = (f_{x_1}, \cdots, f_{x_n})$, $f_\xi = (f_{\xi_1}, \cdots, f_{\xi_n})$ and similarly for the*

Remark 4.17 It is interesting to try to understand, formally, the relationship between the last two theorems, for example in the case where $f = f(x, u, \xi) = f(\xi)$. The coefficients $a_{ij}(x)$ and the function v in Theorem 4.16 are, respectively, $f_{\xi_i \xi_j}(\nabla u(x))$ and u_{x_i} in Theorem 4.14. The fact that $v = u_{x_i} \in W^{1,2}$ is proved by the method of difference quotients presented in Theorem 4.7. This approach is implemented in Exercise 4.7.1.

The two preceding theorems do not generalize to the vectorial case $u : \Omega \subset \mathbb{R}^n \to \mathbb{R}^N$, with $n, N > 1$. In this case only partial regularity can, in general, be proved. We give here an example (cf. Giusti-Miranda [54]) of such a phenomenon.

Example 4.18 Let $n \geq 3$, $\Omega \subset \mathbb{R}^n$ be the unit ball and $u_0(x) = x$. Let $\xi = \left(\xi_i^j \right) \in \mathbb{R}^{n \times n}$ (if $\xi = \nabla u$, then $\xi_i^j = \partial u^j / \partial x_i$) and

$$f(x, u, \xi) = f(u, \xi) = \sum_{i,j=1}^n \left(\xi_i^j \right)^2 + \left[\sum_{i,j=1}^n \left(\delta_{ij} + \frac{4}{n-2} \frac{u^i u^j}{1 + |u|^2} \right) \xi_i^j \right]^2$$

where δ_{ij} is the Kronecker symbol (i.e., $\delta_{ij} = 0$ if $i \neq j$ and $\delta_{ij} = 1$ if $i = j$). Let

$$(P) \quad \inf \left\{ I(u) = \int_\Omega f(u(x), \nabla u(x)) \, dx : u \in u_0 + W_0^{1,2}(\Omega; \mathbb{R}^n) \right\}.$$

(i) It turns out (cf. Exercise 4.7.2) that $\overline{u}(x) = x / |x|$ is a solution of the weak form of the Euler-Lagrange equation associated to (P).

(ii) Moreover it can be shown (cf. [54]) that, for n sufficiently large, $\overline{u}$ is the unique minimizer of (P).

4.7.1 Exercises

Exercise 4.7.1 Let $\Omega \subset \mathbb{R}^n$ be a bounded open set with Lipschitz boundary, $u_0 \in W^{1,2}(\Omega)$ and

$$(P) \quad \inf \left\{ I(u) = \int_\Omega f(\nabla u(x)) \, dx : u \in u_0 + W_0^{1,2}(\Omega) \right\}$$

where $f \in C^2(\mathbb{R}^n)$ is convex and there exist $\gamma_1, \cdots, \gamma_5 > 0$ such that

$$\gamma_1 |\xi|^2 - \gamma_2 \leq f(\xi) \leq \gamma_3 \left(|\xi|^2 + 1 \right), \; \forall \xi \in \mathbb{R}^n$$

$$\left| \frac{\partial^2 f}{\partial \xi_i \partial \xi_j} \right| \leq \gamma_4, \; \forall \xi \in \mathbb{R}^n \; \text{and} \; \forall i, j = 1, \cdots, n$$

As already seen in Chapter 3, even though the function f is strictly convex and $f(\xi) \geq |\xi|^p$ with $p = 1$, we cannot use the direct methods of the calculus of variations, since we are led to work, because of the coercivity condition $f(\xi) \geq |\xi|$, in the non-reflexive space $W^{1,1}(\Omega)$. In fact, in general, there is no minimizer of (P) in $u_0 + W_0^{1,1}(\Omega)$. We therefore need a different approach to deal with this problem.

Before going further we write the associated Euler-Lagrange equation to (P)

$$(E) \quad \operatorname{div}\left[\frac{\nabla u}{\sqrt{1 + |\nabla u|^2}}\right] = \sum_{i=1}^{n} \frac{\partial}{\partial x_i}\left[\frac{u_{x_i}}{\sqrt{1 + |\nabla u|^2}}\right] = 0$$

or equivalently

$$(E) \quad Mu \equiv \left(1 + |\nabla u|^2\right)\Delta u - \sum_{i,j=1}^{n} u_{x_i} u_{x_j} u_{x_i x_j} = 0.$$

The last equation is known as the *minimal surface equation*. If $n = 2$ and $u = u(x,y)$, it reads as

$$Mu = \left(1 + u_y^2\right)u_{xx} - 2u_x u_y u_{xy} + \left(1 + u_x^2\right)u_{yy} = 0.$$

Therefore any $C^2(\overline{\Omega})$ minimizer of (P) should satisfy the equation (E) and conversely, since the integrand f is convex. Moreover, since f is strictly convex, the minimizer, if it exists, is unique. The equation (E) is equivalent (see Section 5.2) to the fact that the *mean curvature* of Σ, denoted by H, vanishes everywhere.

It is clear that the above problem is, geometrically, too restrictive. Indeed if any surface can be locally represented as a graph of a function (i.e. a non-parametric surface), it is not the case globally. We are therefore led to consider more general ones known as *parametric surfaces*. These are sets $\Sigma \subset \mathbb{R}^{n+1}$ so that there exist a domain (i.e. an open and connected set) $\Omega \subset \mathbb{R}^n$ and a map $v : \overline{\Omega} \to \mathbb{R}^{n+1}$ such that

$$\Sigma = v\left(\overline{\Omega}\right) = \left\{v(x) : x \in \overline{\Omega}\right\}.$$

For example, when $n = 2$ and $v = v(x,y) \in \mathbb{R}^3$, if we denote by $v_x \times v_y$ the normal to the surface (where $a \times b$ stands for the vectorial product of $a, b \in \mathbb{R}^3$ and $v_x = \partial v/\partial x$, $v_y = \partial v/\partial y$) we find that the area is given by

$$\text{Area}(\Sigma) = J(v) = \iint_\Omega |v_x \times v_y| \, dx dy.$$

More generally if $n \geq 2$, we define (cf. Theorem 4.4.10 in Morrey [79])

$$g(\nabla v) = \left[\sum_{i=1}^{n+1}\left(\frac{\partial\left(v^1, \cdots, v^{i-1}, v^{i+1}, \cdots, v^{n+1}\right)}{\partial(x_1, \cdots, x_n)}\right)^2\right]^{1/2}$$

problem (the name was given after the theoretical and experimental work of the physicist Plateau) was solved in 1930 simultaneously and independently by Douglas and Rado. One of the first two Fields medals was awarded to Douglas in 1936 for having solved the problem. Before that many mathematicians have contributed to the study of the problem: Ampère, Beltrami, Bernstein, Bonnet, Catalan, Darboux, Enneper, Haar, Korn, Legendre, Lie, Meusnier, Monge, Müntz, Riemann, H.A. Schwarz, Serret, Weierstrass, Weingarten and others. Immediately after the work of Douglas and Rado, we can quote Courant, Mac Shane, Morrey, Morse, Tonelli and many others since then. It is still a very active field.

We conclude this introduction with some comments on the bibliography. We should first point out that we give many results without proofs and the ones that are given are only sketched. It is therefore indispensable in this chapter, even more than in the others, to refer to the bibliography. There are several excellent books but, due to the nature of the subject, they are difficult to read. The most complete to which we constantly refer are those of Dierkes-Hildebrandt-Küster-Wohlrab [39] and Nitsche [82]. As a matter of introduction, interesting for a general audience, one can consult Hildebrandt-Tromba [61]. We also refer to the monographs of Almgren [4], Courant [24], Federer [46], Gilbarg-Trudinger [51] (for the nonparametric surfaces), Giusti [52], Morrey [79], Osserman [84] and Struwe [96].

5.2   Generalities about surfaces

We now introduce the different types of surfaces that we consider. We essentially limit ourselves to surfaces of $\mathbb{R}^3$, although in some instances we give some generalizations to $\mathbb{R}^{n+1}$. Besides the references that we already mentioned, one can consult books of differential geometry such as that of Hsiung [65].

Definition 5.1 *(i) A set $\Sigma \subset \mathbb{R}^3$ is called a* parametric surface *(or more simply a surface) if there exist a domain (i.e. an open and connected set) $\Omega \subset \mathbb{R}^2$ and a (non-constant) continuous map $v : \overline{\Omega} \to \mathbb{R}^3$ such that*

$$\Sigma = v\left(\overline{\Omega}\right) = \left\{ v\left(x,y\right) \in \mathbb{R}^3 : (x,y) \in \overline{\Omega} \right\}.$$

(ii) We say that Σ is a nonparametric surface *if*

$$\Sigma = \left\{ v\left(x,y\right) = (x,y,u\left(x,y\right)) \in \mathbb{R}^3 : (x,y) \in \overline{\Omega} \right\}$$

with $u : \overline{\Omega} \to \mathbb{R}$ continuous and where $\Omega \subset \mathbb{R}^2$ is a domain.

(iii) A parametric surface is said to be regular *of class C^m, (m $\geq$ 1 an integer) if, in addition, $v \in C^m\left(\Omega; \mathbb{R}^3\right)$ and*

$$v_x \times v_y \neq 0 \quad \text{for every } (x,y) \in \Omega$$

(iii) The principal curvatures, k_1 and k_2, are defined as

$$k_1 = H + \sqrt{H^2 - K} \quad and \quad k_2 = H - \sqrt{H^2 - K}$$

so that

$$H = \frac{k_1 + k_2}{2} \quad and \quad K = k_1 k_2.$$

Remark 5.4 (i) We always have $H^2 \geq K$.

(ii) For a nonparametric surface $v(x, y) = (x, y, \ u(x, y))$, we have

$$E = 1 + u_x^2, \ F = u_x u_y, \ G = 1 + u_y^2, \ EG - F^2 = 1 + u_x^2 + u_y^2$$

$$e_3 = \frac{(-u_x, -u_y, 1)}{\sqrt{1 + u_x^2 + u_y^2}}, \ L = \frac{u_{xx}}{\sqrt{1 + u_x^2 + u_y^2}},$$

$$M = \frac{u_{xy}}{\sqrt{1 + u_x^2 + u_y^2}}, \ N = \frac{u_{yy}}{\sqrt{1 + u_x^2 + u_y^2}}$$

and hence

$$H = \frac{\left(1 + u_y^2\right) u_{xx} - 2 u_x u_y u_{xy} + \left(1 + u_x^2\right) u_{yy}}{2 \left(1 + u_x^2 + u_y^2\right)^{3/2}} \quad and \quad K = \frac{u_{xx} u_{yy} - u_{xy}^2}{\left(1 + u_x^2 + u_y^2\right)^2}.$$

(iii) For a nonparametric surface in $\mathbb{R}^{n+1}$ given by $x_{n+1} = u(x_1, \cdots, x_n)$, we have that the mean curvature is defined by (cf. (A.14) in Gilbarg-Trudinger [51])

$$H = \frac{1}{n} \sum_{i=1}^{n} \frac{\partial}{\partial x_i} \left[\frac{u_{x_i}}{\sqrt{1 + |\nabla u|^2}} \right] = \frac{\left(1 + |\nabla u|^2\right) \Delta u - \sum_{i,j=1}^{n} u_{x_i} u_{x_j} u_{x_i x_j}}{n \left(1 + |\nabla u|^2\right)^{3/2}}.$$

In terms of the operator M defined in the introduction of the present chapter, we can write

$$Mu = n \left(1 + |\nabla u|^2\right)^{3/2} H.$$

(iv) Note that we always have (see Exercise 5.2.1)

$$|v_x \times v_y| = \sqrt{EG - F^2}.$$

We are now in a position to define the notion of minimal surface.

Proposition 5.11 *The only regular minimal surfaces of revolution of the form*

$$v(x, y) = (x, w(x) \cos y, w(x) \sin y),$$

are the catenoids, *i.e.*

$$w(x) = \lambda \cosh \frac{x + \mu}{\lambda}$$

where $\lambda \neq 0$ *and* μ *are constants.*

Proof. We have to prove that Σ given parametrically by v is minimal if and only if

$$w(x) = \lambda \cosh((x + \mu)/\lambda).$$

Observe first that

$$v_x = (1, w' \cos y, w' \sin y), \; v_y = (0, -w \sin y, w \cos y)$$

$$E = 1 + w'^2, \; F = 0, \; G = w^2$$

$$v_x \times v_y = w(w', -\cos y, -\sin y), \; e_3 = \frac{w}{|w|} \frac{(w', -\cos y, -\sin y)}{\sqrt{1 + w'^2}}$$

$$v_{xx} = w''(0, \cos y, \sin y), \; v_{xy} = w'(0, -\sin y, \cos y), \; v_{yy} = -w(0, \cos y, \sin y)$$

$$L = \frac{w}{|w|} \frac{-w''}{\sqrt{1 + w'^2}}, \; M = 0, \; N = \frac{|w|}{\sqrt{1 + w'^2}}.$$

Since Σ is a regular surface, we must have $|w| > 0$ (because $|v_x \times v_y|^2 = EG - F^2 > 0$). We therefore deduce that

$$H = 0 \Leftrightarrow EN + GL = 0 \Leftrightarrow |w| \left(ww'' - (1 + w'^2)\right) = 0$$

and thus $H = 0$ is equivalent to

$$ww'' = 1 + w'^2. \tag{5.1}$$

Any solution of the differential equation necessarily satisfies

$$\frac{d}{dx} \left[\frac{w(x)}{\sqrt{1 + w'^2(x)}} \right] = 0.$$

The solution of this last differential equation (see the solution to Exercise 5.2.3) being either $w \equiv$ constant (which however does not satisfy (5.1)) or of the form

$$w(x) = \lambda \cosh \left(\frac{x + \mu}{\lambda} \right),$$

we have the result. ∎

We now turn our attention to the relationship between minimal surfaces and surfaces of minimal area.

and hence

$$\frac{d}{d\epsilon} I \left(\overline{u} + \epsilon\varphi\right)\bigg|_{\epsilon=0} = \iint_\Omega \frac{\overline{u}_x \varphi_x + \overline{u}_y \varphi_y}{\sqrt{1 + \overline{u}_x^2 + \overline{u}_y^2}} \, dxdy = 0, \ \forall \varphi \in C_0^\infty \left(\Omega\right).$$

Since $\overline{u} \in C^2 \left(\overline{\Omega}\right)$ we have, after integration by parts and using the fundamental lemma of the calculus of variations (Theorem 1.24),

$$\frac{\partial}{\partial x}\left[\frac{\overline{u}_x}{\sqrt{1 + \overline{u}_x^2 + \overline{u}_y^2}}\right] + \frac{\partial}{\partial y}\left[\frac{\overline{u}_y}{\sqrt{1 + \overline{u}_x^2 + \overline{u}_y^2}}\right] = 0 \text{ in } \overline{\Omega} \qquad (5.2)$$

or equivalently

$$M\overline{u} = \left(1 + \overline{u}_y^2\right)\overline{u}_{xx} - 2\overline{u}_x \overline{u}_y \overline{u}_{xy} + \left(1 + \overline{u}_x^2\right)\overline{u}_{yy} = 0 \text{ in } \overline{\Omega}. \qquad (5.3)$$

This just asserts that $H = 0$ and hence

$$\Sigma_{\overline{u}} = \left\{(x, y, \overline{u}\left(x, y\right)) : (x, y) \in \overline{\Omega}\right\}$$

is a minimal surface.

(i) $\Rightarrow$ (ii). We start by noting that the function

$$\xi \to f\left(\xi\right) = \sqrt{1 + |\xi|^2}$$

where $\xi \in \mathbb{R}^2$, is strictly convex. So let

$$\Sigma_{\overline{u}} = \left\{(x, y, \overline{u}\left(x, y\right)) : (x, y) \in \overline{\Omega}\right\}$$

be a minimal surface. Since $H = 0$, we have that $\overline{u}$ satisfies (5.2) or (5.3). Let

$$\Sigma_u = \left\{(x, y, u\left(x, y\right)) : (x, y) \in \overline{\Omega}\right\}$$

with $u \in C^2 \left(\overline{\Omega}\right)$ and $u = \overline{u}$ on $\partial\Omega$. We want to show that $I\left(\overline{u}\right) \leq I\left(u\right)$. Since f is convex, we have

$$f\left(\xi\right) \geq f\left(\eta\right) + \langle \nabla f\left(\eta\right); \xi - \eta \rangle, \ \forall \xi, \eta \in \mathbb{R}^2$$

and hence

$$f\left(u_x, u_y\right) \geq f\left(\overline{u}_x, \overline{u}_y\right) + \frac{1}{\sqrt{1 + \overline{u}_x^2 + \overline{u}_y^2}} \langle (\overline{u}_x, \overline{u}_y); (u_x - \overline{u}_x, u_y - \overline{u}_y) \rangle.$$

Integrating the above inequality and appealing to (5.2) and to the fact that $u = \overline{u}$ on $\partial\Omega$ we readily obtain the result.

To conclude we point out the deep relationship between isothermal coordinates of minimal surfaces and harmonic functions (see also Theorem 5.15) which is one of the basic facts in the proof of Douglas.

Theorem 5.17 *Let*

$$\Sigma = \left\{ v\left(x, y\right) \in \mathbb{R}^3 : \left(x, y\right) \in \overline{\Omega} \right\}$$

be a regular surface (i.e. $v_x \times v_y \neq 0$) of class C^2 globally parametrized by isothermal coordinates; then

$$\Sigma \text{ is a minimal surface} \quad \Leftrightarrow \quad \Delta v = 0 \text{ (i.e. } \Delta v^1 = \Delta v^2 = \Delta v^3 = 0\text{)}.$$

Proof. We will show that if $E = G = |v_x|^2 = |v_y|^2$ and $F = 0$, then

$$\Delta v = 2EH e_3 = 2H\, v_x \times v_y \qquad (5.4)$$

where H is the mean curvature and $e_3 = (v_x \times v_y) / |v_x \times v_y|$. The result readily follows, since $v_x \times v_y \neq 0$.

Since $E = G$ and $F = 0$, we have

$$H = \frac{L + N}{2E} \quad \Rightarrow \quad L + N = 2EH. \qquad (5.5)$$

We next prove that $\langle v_x; \Delta v \rangle = \langle v_y; \Delta v \rangle = 0$. Using the equations $E = G$ and $F = 0$, we have, after differentiation of the first one by x and the second one by y,

$$\langle v_x; v_{xx} \rangle = \langle v_y; v_{xy} \rangle \quad \text{and} \quad \langle v_x; v_{yy} \rangle + \langle v_y; v_{xy} \rangle = 0.$$

This leads, as wished, to $\langle v_x; \Delta v \rangle = 0$ and in a similar way to $\langle v_y; \Delta v \rangle = 0$. Therefore Δv is orthogonal to v_x and v_y and thus parallel to e_3, which means that there exists $a \in \mathbb{R}$ so that $\Delta v = a e_3$. We then deduce that

$$a = \langle e_3; \Delta v \rangle = \langle e_3; v_{xx} \rangle + \langle e_3; v_{yy} \rangle = L + N. \qquad (5.6)$$

Combining (5.5) and (5.6), we immediately get (5.4) and the theorem then follows. ∎

5.2.1 Exercises

Exercise 5.2.1 **(i)** Let $a, b, c \in \mathbb{R}^3$ show that

$$|a \times b|^2 = |a|^2 |b|^2 - (\langle a; b \rangle)^2$$

$$(a \times b) \times c = \langle a; c \rangle\, b - \langle b; c \rangle\, a.$$

and $\Gamma \subset \mathbb{R}^3$ be a rectifiable (i.e. of finite length) simple closed curve. Let $w_i \in \partial\Omega$ $(w_i \neq w_j)$ and $p_i \in \Gamma$ $(p_i \neq p_j)$ $i = 1, 2, 3$ be fixed. The set of admissible surfaces is then

$$\mathcal{S} = \left\{ \begin{array}{ll} \Sigma = v\left(\overline{\Omega}\right) \text{ where } & v : \overline{\Omega} \to \Sigma \subset \mathbb{R}^3 \text{ so that} \\ & (S_1) \quad v \in \mathcal{M}\left(\overline{\Omega}\right) = C^0\left(\overline{\Omega}; \mathbb{R}^3\right) \cap W^{1,2}\left(\Omega; \mathbb{R}^3\right) \\ & (S_2) \quad v : \partial\Omega \to \Gamma \text{ is weakly monotonic and onto} \\ & (S_3) \quad v\left(w_i\right) = p_i, \quad i = 1, 2, 3 \end{array} \right\}.$$

Remark 5.18 (i) The set of admissible surfaces is then the set of parametric surfaces of the type of the disk with parametrization in $\mathcal{M}\left(\overline{\Omega}\right)$. The condition weakly monotonic in (S_2) means that we allow the map v to be constant on some parts of $\partial\Omega$; thus v is not necessarily a homeomorphism of $\partial\Omega$ onto Γ. However the minimizer of the theorem has the property to map the boundary $\partial\Omega$ topologically onto the simple closed curve Γ. The condition (S_3) may appear a little strange, it helps us to get compactness (see the proof below).

(ii) A first natural question is to ask if $\mathcal{S}$ is non-empty. If the simple closed curve Γ is rectifiable then $\mathcal{S} \neq \emptyset$ (see for more details Dierkes-Hildebrandt-Küster-Wohlrab [39] pages 232-234 and Nitsche [82] pages 253-257).

(iii) Recall from the preceding section that for $\Sigma \in \mathcal{S}$ we have

$$\text{Area}\left(\Sigma\right) = J\left(v\right) = \iint_\Omega \left|v_x \times v_y\right| dx dy.$$

The main result of this chapter is then

Theorem 5.19 *Under the above hypotheses there exists $\Sigma_0 \in \mathcal{S}$ so that*

$$\text{Area}\left(\Sigma_0\right) \leq \text{Area}\left(\Sigma\right), \quad \forall \Sigma \in \mathcal{S}.$$

Moreover there exists $\overline{v}$ satisfying (S_1), (S_2) and (S_3), such that $\Sigma_0 = \overline{v}\left(\overline{\Omega}\right)$ and

(i) $\overline{v} \in C^\infty\left(\Omega; \mathbb{R}^3\right)$ *with $\Delta\overline{v} = 0$ in Ω,*

(ii) $E = \left|\overline{v}_x\right|^2 = G = \left|\overline{v}_y\right|^2$ *and $F = \langle\overline{v}_x; \overline{v}_y\rangle = 0$.*

(iii) $\overline{v}$ *maps the boundary $\partial\Omega$ topologically onto the simple closed curve Γ.*

Remark 5.20 (i) The theorem asserts that Σ_0 is of minimal area. To solve completely Plateau problem, we must still prove that Σ_0 is a regular surface (i.e. $\overline{v}_x \times \overline{v}_y \neq 0$ everywhere); we will then be able to apply Theorem 5.12 to conclude. We mention in the next section some results concerning this problem. We also have a regularity result, namely that $\overline{v}$ is C^∞ and harmonic, as well as a choice of isothermal coordinates ($E = G$ and $F = 0$).

Such a $\widetilde{v}_\nu$ exists and its components are harmonic (cf. Chapter 3). Combining (5.9) and (5.10), we still have

$$D(\widetilde{v}_\nu) \to d.$$

Without the hypotheses (S_2), (S_3), this new sequence $\{\widetilde{v}_\nu\}$ does not converge either. The condition (S_3) is important, since (see Exercise 5.3.1) Dirichlet integral is invariant under any conformal transformation from Ω onto Ω; (S_3) allows to select a unique one. The hypothesis (S_2) and the Courant-Lebesgue lemma imply that $\{\widetilde{v}_\nu\}$ is a sequence of equicontinuous functions (see Courant [24] page 103, Dierkes-Hildebrandt-Küster-Wohlrab [39] pages 235-237 or Nitsche [82] page 257). It follows from Ascoli-Arzelà theorem (Theorem 1.3) that, up to a subsequence,

$$\widetilde{v}_{\nu_k} \to \overline{v} \text{ uniformly.}$$

Harnack theorem (see, for example, Gilbarg-Trudinger [51], page 21), a classical property of harmonic functions, implies that $\overline{v}$ is harmonic, satisfies (S_1), (S_2), (S_3) and

$$D(\overline{v}) = d \quad \text{with} \quad \Delta \overline{v} = 0 \text{ in } \Omega. \tag{5.11}$$

Step 3. We next show that this map $\overline{v}$ verifies also $E = G$ (i.e. $|\overline{v}_x|^2 = |\overline{v}_y|^2$) and $F = 0$ (i.e. $\langle \overline{v}_x; \overline{v}_y \rangle = 0$), which in particular implies that

$$\text{Area}\left(\overline{v}(\overline{\Omega})\right) = D(\overline{v}).$$

We use, in order to establish this fact, the technique of variations of the independent variables that we have already encountered in Section 2.3, when deriving the second form of the Euler-Lagrange equation. Since the proof of this step is lengthy, we subdivide it into three substeps.

Step 3.1. Let $\lambda, \mu \in C^\infty(\overline{\Omega})$, to be chosen later, and let $\epsilon \in \mathbb{R}$ be sufficiently small so that the map

$$\begin{pmatrix} x' \\ y' \end{pmatrix} = \varphi^\epsilon(x,y) = \begin{pmatrix} \varphi_1^\epsilon(x,y) \\ \varphi_2^\epsilon(x,y) \end{pmatrix} = \begin{pmatrix} x + \epsilon\lambda(x,y) \\ y + \epsilon\mu(x,y) \end{pmatrix}$$

is a diffeomorphism from $\overline{\Omega}$ onto a simply connected domain $\overline{\Omega}^\epsilon = \varphi^\epsilon(\Omega)$. We denote its inverse by ψ^ϵ and we find that

$$\begin{pmatrix} x \\ y \end{pmatrix} = \psi^\epsilon(x',y') = \begin{pmatrix} \psi_1^\epsilon(x',y') \\ \psi_2^\epsilon(x',y') \end{pmatrix} = \begin{pmatrix} x' - \epsilon\lambda(x',y') + o(\epsilon) \\ y' - \epsilon\mu(x',y') + o(\epsilon) \end{pmatrix}$$

where $o(t)$ stands for a function $f = f(t)$ so that $f(t)/t$ tends to 0 as t tends to 0. We therefore have

$$\varphi^\epsilon(\psi^\epsilon(x',y')) = (x',y') \quad \text{and} \quad \psi^\epsilon(\varphi^\epsilon(x,y)) = (x,y)$$

Since $\overline{v} \in \mathcal{S}$, we deduce that $v^\epsilon \in \mathcal{S}$. Therefore using the conformal invariance of the Dirichlet integral (see Exercise 5.3.1), we find that

$$
\begin{aligned}
D\left(v^\epsilon\right) &= \frac{1}{2} \iint_\Omega \left[\left|v_x^\epsilon\left(x,y\right)\right|^2 + \left|v_y^\epsilon\left(x,y\right)\right|^2\right] dxdy \\
&= \frac{1}{2} \iint_{\Omega^\epsilon} \left[\left|u_{x'}^\epsilon\left(x',y'\right)\right|^2 + \left|u_{y'}^\epsilon\left(x',y'\right)\right|^2\right] dx'dy'
\end{aligned}
$$

which combined with (5.13) leads to

$$
\begin{aligned}
D\left(v^\epsilon\right) &= D\left(\overline{v}\right) - \frac{\epsilon}{2} \iint_\Omega \left[\left(\left|\overline{v}_x\right|^2 - \left|\overline{v}_y\right|^2\right)\left(\lambda_x - \mu_y\right)\right] dxdy \\
&\quad - \epsilon \iint_\Omega \left[\langle\overline{v}_x; \overline{v}_y\rangle\left(\lambda_y + \mu_x\right)\right] dxdy + o\left(\epsilon\right).
\end{aligned}
$$

Since $v^\epsilon, \overline{v} \in \mathcal{S}$ and $\overline{v}$ is a minimizer of the Dirichlet integral, we find that

$$
\iint_\Omega \left[\left(\left|\overline{v}_x\right|^2 - \left|\overline{v}_y\right|^2\right)\left(\lambda_x - \mu_y\right) + 2\langle\overline{v}_x; \overline{v}_y\rangle\left(\lambda_y + \mu_x\right)\right] dxdy = 0. \qquad (5.14)
$$

Step 3.3. We finally choose in an appropriate way the functions $\lambda, \mu \in C^\infty\left(\overline{\Omega}\right)$ that appeared in the previous steps. We let $\sigma, \tau \in C_0^\infty\left(\Omega\right)$ be arbitrary, we then choose λ and μ so that

$$
\begin{cases}
\lambda_x - \mu_y = \sigma \\
\lambda_y + \mu_x = \tau
\end{cases}
$$

(this is always possible; find first λ satisfying

$$
\Delta\lambda = \sigma_x + \tau_y
$$

then choose μ such that $\left(\mu_x, \mu_y\right) = \left(\tau - \lambda_y, \lambda_x - \sigma\right))$. Returning to (5.14) we find

$$
\iint_\Omega \left\{\left(E - G\right)\sigma + 2F\tau\right\} dxdy = 0, \ \forall \sigma, \tau \in C_0^\infty\left(\Omega\right).
$$

The fundamental lemma of the calculus of variations (Theorem 1.24) implies then $E = G$ and $F = 0$. Thus, up to the condition $\overline{v}_x \times \overline{v}_y \neq 0$, Plateau problem is solved (cf. Theorem 5.17), since $\Delta\overline{v} = 0$, according to (5.11). We have thus found $\overline{v}$ satisfying (i) and (ii) in the statements of the theorem.

We still have to prove (iii) in the statement of the theorem. However this follows easily from (i), (ii) and (S_2) of the theorem, cf. Dierkes-Hildebrandt-Küster-Wohlrab [39] page 248.

Theorem 5.21 *If* Γ *is a simple closed curve of class* $C^{k,\alpha}$ *(with* $k \geq 1$ *an integer and* $0 < \alpha < 1$ *) then every solution of Plateau problem (i.e. a minimal surface whose boundary is* Γ *) admits a* $C^{k,\alpha}\left(\overline{\Omega}\right)$ *parametrization.*

However the most important regularity result concerns the existence of a *regular surface* (i.e. with $v_x \times v_y \neq 0$) which solves Plateau problem. We have seen in Section 5.3, that the method of Douglas does not answer this question. A result in this direction is the following (see Nitsche [82] page 334).

Theorem 5.22 *(i) If* Γ *is an analytical simple closed curve and if its total curvature does not exceed* 4π, *then any solution of Plateau problem is a regular minimal surface.*

(ii) If a solution of Plateau problem is of minimal area then the result remains true without any hypothesis on the total curvature of Γ.

Remark 5.23 The second part of the theorem allows, a posteriori, to assert that the solution found in Section 5.3 is a regular minimal surface (i.e. $\overline{v}_x \times \overline{v}_y \neq 0$), provided Γ is analytical.

We now turn our attention to the problem of uniqueness of minimal surfaces. Recall first (Theorem 5.12) that we have uniqueness when restricted to nonparametric surfaces. For general surfaces we have the following uniqueness result (see Nitsche [82] page 351).

Theorem 5.24 *Let* Γ *be an analytical simple closed curve with total curvature not exceeding* 4π, *then Plateau problem has a unique solution.*

We now give a non-uniqueness result (for more details we refer to Dierkes-Hildebrandt-Küster-Wohlrab [39] and to Nitsche [82]).

Example 5.25 (Enneper surface) (see Example 5.10). Let $r \in \left(1, \sqrt{3}\right)$ and

$$\Gamma_r = \left\{ \left(r \cos \theta - \frac{1}{3}r^3 \cos 3\theta, -r \sin \theta - \frac{1}{3}r^3 \sin 3\theta, r^2 \cos 2\theta \right) : \theta \in [0, 2\pi) \right\}.$$

We have seen (Example 5.10) that

$$\Sigma_r = \left\{ \left(rx + r^3xy^2 - \frac{r^3}{3}x^3, -ry - r^3x^2y + \frac{r^3}{3}y^3, r^2\left(x^2 - y^2\right) \right) : x^2 + y^2 \leq 1 \right\}$$

is a minimal surface and that $\partial \Sigma_r = \Gamma_r$. It is possible to show (cf. Nitsche [82] page 338) that Σ_r is not of minimal area if $r \in \left(1, \sqrt{3}\right)$; therefore it is distinct from the one found in Theorem 5.19.

with $a \in \mathbb{R}^n$ and $b \in \mathbb{R}$, which in geometrical terms represents a hyperplane, is a solution of the equation. The question is to know if this is the only one.

In the case $n = 2$, Bernstein has shown that, indeed, this is the only C^2 solution (the result is known as Bernstein theorem). Since then several authors found different proofs of this theorem. The extension to higher dimensions is however much harder. De Giorgi extended the result to the case $n = 3$, Almgren to the case $n = 4$ and Simons to $n = 5, 6, 7$. In 1969, Bombieri, De Giorgi and Giusti proved that when $n \geq 8$, there exists a nonlinear $u \in C^2(\mathbb{R}^n)$ (and hence the surface is not a hyperplane) satisfying equation (E). For more details on Bernstein problem, see Giusti [52] Chapter 17 and Nitsche [82] pages 123-124 and 429-430.

We now return to our problem in a bounded domain. We start by quoting a result of Jenkins and Serrin; for a proof see Gilbarg-Trudinger [51] page 297.

Theorem 5.26 (Jenkins-Serrin) *Let $\Omega \subset \mathbb{R}^n$ be a bounded domain with $C^{2,\alpha}$, $0 < \alpha < 1$, boundary and let $u_0 \in C^{2,\alpha}(\overline{\Omega})$. The problem*

$$\begin{cases} Mu = 0 & in \ \Omega \\ u = u_0 & on \ \partial\Omega \end{cases}$$

has a solution for every u_0 if and only if the mean curvature of $\partial\Omega$ is everywhere non-negative.

Remark 5.27 (i) We now briefly mention a related result due to Finn and Osserman. It roughly says that if Ω is a non-convex domain, there exists a continuous u_0 so that the problem

$$\begin{cases} Mu = 0 & in \ \Omega \\ u = u_0 & on \ \partial\Omega \end{cases}$$

has no C^2 solution. Such a u_0 can even have arbitrarily small norm $\|u_0\|_{C^0}$.

(ii) The above theorem follows several earlier works that started with Bernstein (see Nitsche [82] pages 352-358).

We end the present chapter with a simple theorem whose ideas contained in the proof are used in several different problems of partial differential equations.

Theorem 5.28 (Korn-Müntz) *Let $\Omega \subset \mathbb{R}^n$ be a bounded domain with $C^{2,\alpha}$, $0 < \alpha < 1$, boundary and consider the problem*

$$\begin{cases} Mu = \left(1 + |\nabla u|^2\right) \Delta u - \sum_{i,j=1}^{n} u_{x_i} u_{x_j} u_{x_i x_j} = 0 & in \ \Omega \\ u = u_0 & on \ \partial\Omega. \end{cases}$$

we deduce that (5.18) holds from Proposition 1.9 and its proof (cf. Exercise 1.2.2).

Step 3. We are now in a position to show the theorem. We define a sequence $\{u_\nu\}_{\nu=1}^{\infty}$ of $C^{2,\alpha}(\overline{\Omega})$ functions in the following way

$$\begin{cases} \Delta u_1 = 0 & \text{in } \Omega \\ u_1 = u_0 & \text{on } \partial\Omega \end{cases} \tag{5.19}$$

and by induction

$$\begin{cases} \Delta u_{\nu+1} = N(u_\nu) & \text{in } \Omega \\ u_{\nu+1} = u_0 & \text{on } \partial\Omega. \end{cases} \tag{5.20}$$

The previous estimates will allow us (cf. Step 4) to deduce that for $\|u_0\|_{C^{2,\alpha}} \leq \epsilon$, ϵ to be determined, we have

$$\|u_{\nu+1} - u_\nu\|_{C^{2,\alpha}} \leq K \|u_\nu - u_{\nu-1}\|_{C^{2,\alpha}} \tag{5.21}$$

for some $K < 1$. If (5.21) is true, we obtain

$$\|u_{\nu+1} - u_\nu\|_{C^{2,\alpha}} \leq K^\nu \|u_1 - u_0\|_{C^{2,\alpha}}.$$

Therefore if $\nu > \mu$, we have

$$\|u_\nu - u_\mu\|_{C^{2,\alpha}} \leq \left(\sum_{s=\mu}^{\nu-1} K^s\right) \|u_1 - u_0\|_{C^{2,\alpha}} \leq \frac{K^\mu}{1-K} \|u_1 - u_0\|_{C^{2,\alpha}}.$$

This readily implies that $\{u_\nu\}$ is a Cauchy sequence in $C^{2,\alpha}$ and thus it converges to a certain u in $C^{2,\alpha}$. We moreover have

$$\begin{cases} \Delta u = N(u) & \text{in } \Omega \\ u = u_0 & \text{on } \partial\Omega \end{cases}$$

which is the claimed result.

Step 4. We now establish (5.21), which amounts to find the appropriate $\epsilon > 0$. We start by choosing $0 < K < 1$ and we then choose $\epsilon > 0$ sufficiently small so that

$$2\gamma^2\epsilon\left(1 + \frac{\gamma^4\epsilon^2}{1-K}\right) \leq \sqrt{K} \tag{5.22}$$

where γ is the constant appearing in Step 1 and Step 2 (we can consider, without loss of generality, that they are the same).

We therefore only need to show that if $\|u_0\|_{C^{2,\alpha}} \leq \epsilon$, we have indeed (5.21). Note that for every $\nu \geq 2$, we have

$$\begin{cases} \Delta(u_{\nu+1} - u_\nu) = N(u_\nu) - N(u_{\nu-1}) & \text{in } \Omega \\ u_{\nu+1} - u_\nu = 0 & \text{on } \partial\Omega. \end{cases}$$

5.5.1   Exercise

Exercise 5.5.1 Let $u \in C^2(\mathbb{R}^2)$, $u = u(x,y)$, be a solution of the minimal surface equation

$$Mu = \left(1 + u_y^2\right) u_{xx} - 2u_x u_y u_{xy} + \left(1 + u_x^2\right) u_{yy} = 0.$$

Show that there exists a convex function $\varphi \in C^2(\mathbb{R}^2)$, so that

$$\varphi_{xx} = \frac{1 + u_x^2}{\sqrt{1 + u_x^2 + u_y^2}}, \ \varphi_{xy} = \frac{u_x u_y}{\sqrt{1 + u_x^2 + u_y^2}}, \ \varphi_{yy} = \frac{1 + u_y^2}{\sqrt{1 + u_x^2 + u_y^2}}.$$

Deduce that

$$\varphi_{xx}\varphi_{yy} - \varphi_{xy}^2 = 1.$$

form (cf. Poincaré-Wirtinger inequality). This inequality reads as

$$\int_{-1}^{1} u'^2 \, dx \geq \pi^2 \int_{-1}^{1} u^2 \, dx, \ \forall u \in X$$

where

$$X = \left\{ u \in W^{1,2}(-1,1) : u(-1) = u(1) \text{ and } \int_{-1}^{1} u \, dx = 0 \right\}.$$

It also states that equality holds if and only if $u(x) = \alpha \cos(\pi x) + \beta \sin(\pi x)$, for any $\alpha, \beta \in \mathbb{R}$.

In Section 6.3 we discuss the generalization to $\mathbb{R}^n$, $n \geq 3$, of the isoperimetric inequality. It reads as follows

$$[L(\partial A)]^n - n^n \omega_n [M(A)]^{n-1} \geq 0$$

for every bounded open set $A \subset \mathbb{R}^n$ with sufficiently regular boundary, ∂A; and where ω_n is the measure of the unit ball of $\mathbb{R}^n$, $M(A)$ stands for the measure of A and $L(\partial A)$ for the $(n-1)$ measure of ∂A. Moreover, if A is sufficiently regular (for example, convex), there is equality if and only if A is a ball.

The inequality in higher dimensions is considerably harder to prove; we briefly discuss, in Section 6.3, the main ideas of the proof. When $n = 3$, the first complete proof is the one of H.A. Schwarz. Soon after there were generalizations to higher dimensions and other proofs notably by A. Aleksandrov, Blaschke, Bonnesen, H. Hopf, Liebmann, Minkowski and E. Schmidt.

Finally numerous generalizations of this inequality have been studied in relation to problems of mathematical physics, see Bandle [9], Payne [87] and Polya-Szegö [89] for more references.

There are several articles and books devoted to the subject, we recommend the review article of Osserman [85] and the books by Berger [10], Blaschke [11], Federer [46], Hardy-Littlewood-Polya [58] (for the two dimensional case) and Webster [101]. The book of Hildebrandt-Tromba [61] also has a chapter on this matter.

6.2   The case of dimension 2

We start with the key result for proving the isoperimetric inequality; but before that we introduce the following notation, for any $p \geq 1$,

$$W_{\text{per}}^{1,p}(a,b) = \left\{ u \in W^{1,p}(a,b) : u(a) = u(b) \right\}$$

and

$$C_{\text{per}}^1([a,b]) = \left\{ u \in C^1([a,b]) : u(a) = u(b) \right\}.$$

Proof. An alternative proof, more in the spirit of Example 2.24, is proposed in Exercise 6.2.1. The proof given here is, essentially, the classical proof of Hurwitz. We divide the proof into two steps.

Step 1. We start by proving the theorem under the further restriction that $u \in X \cap C^2[-1,1]$. We express u in Fourier series

$$u(x) = \sum_{n=1}^{\infty} [a_n \cos(n\pi x) + b_n \sin(n\pi x)].$$

Note that there is no constant term since $\int_{-1}^{1} u(x)\, dx = 0$. We know at the same time that

$$u'(x) = \pi \sum_{n=1}^{\infty} [-na_n \sin(n\pi x) + nb_n \cos(n\pi x)].$$

We can now invoke *Parseval formula* to get

$$\int_{-1}^{1} u^2\, dx = \sum_{n=1}^{\infty} (a_n^2 + b_n^2) \quad \text{and} \quad \int_{-1}^{1} u'^2\, dx = \pi^2 \sum_{n=1}^{\infty} (a_n^2 + b_n^2)\, n^2.$$

The desired inequality follows then at once

$$\int_{-1}^{1} u'^2\, dx \geq \pi^2 \int_{-1}^{1} u^2\, dx, \ \forall u \in X \cap C^2.$$

Moreover equality holds if and only if $a_n = b_n = 0$, for every $n \geq 2$. This implies that equality holds if and only if $u(x) = \alpha \cos(\pi x) + \beta \sin(\pi x)$, for any $\alpha, \beta \in \mathbb{R}$, as claimed.

Step 2. We now show that we can remove the restriction $u \in X \cap C^2[-1,1]$. By the usual density argument we can find for every $u \in X$ a sequence $u_\nu \in X \cap C^2[-1,1]$ so that

$$u_\nu \to u \text{ in } W^{1,2}(-1,1).$$

Therefore, for every $\epsilon > 0$, we can find ν sufficiently large so that

$$\int_{-1}^{1} u'^2\, dx \geq \int_{-1}^{1} u_\nu'^2\, dx - \epsilon \quad \text{and} \quad \int_{-1}^{1} u_\nu^2\, dx \geq \int_{-1}^{1} u^2\, dx - \epsilon.$$

Combining these inequalities with Step 1 we find

$$\int_{-1}^{1} u'^2\, dx \geq \pi^2 \int_{-1}^{1} u^2\, dx - (\pi^2 + 1)\, \epsilon.$$

Letting $\epsilon \to 0$ we have indeed obtained the inequality.

and thus
$$f^2 + g^2 = \alpha^2 + \beta^2.$$
We therefore have, setting $r_3 = \sqrt{\alpha^2 + \beta^2}$, that
$$(u(x) - r_1)^2 + (v(x) - r_2)^2 = r_3^2, \quad \forall x \in [-1, 1]$$
as wished. ∎

We are now in a position to prove the isoperimetric inequality in its analytic form; we postpone the discussion of its geometric meaning for later.

Theorem 6.4 (Isoperimetric inequality) *For* $u, v \in W_{per}^{1,1}(a, b)$, *let*

$$L(u, v) = \int_a^b \sqrt{u'^2 + v'^2} \, dx$$

$$M(u, v) = \frac{1}{2} \int_a^b (uv' - vu') \, dx = \int_a^b uv' \, dx.$$

Then

$$[L(u, v)]^2 - 4\pi M(u, v) \geq 0, \quad \forall u, v \in W_{per}^{1,1}(a, b).$$

Moreover, among all $u, v \in C_{per}^1([a, b])$, *equality holds if and only if*

$$(u(x) - r_1)^2 + (v(x) - r_2)^2 = r_3^2, \quad \forall x \in [a, b]$$

where $r_1, r_2, r_3 \in \mathbb{R}$ *are constants.*

Remark 6.5 The uniqueness holds under weaker regularity hypotheses, but we do not discuss here this matter. We, however, point out that the very same proof for the uniqueness is valid for piecewise $C_{per}^1([a, b])$ u and v.

Proof. We divide the proof into two steps.

Step 1. We first prove the theorem under the further restriction that $u, v \in C_{per}^1([a, b])$. We also assume that

$$u'^2(x) + v'^2(x) > 0, \quad \forall x \in [a, b].$$

This hypothesis is unnecessary and can be removed, see Exercise 6.2.3.

We start by reparametrizing the curve by a multiple of its arc length, namely

$$\begin{cases} y = \eta(x) = -1 + \frac{2}{L(u,v)} \int_a^x \sqrt{u'^2 + v'^2} \, dx \\ \varphi(y) = u(\eta^{-1}(y)) \quad \text{and} \quad \psi(y) = v(\eta^{-1}(y)). \end{cases}$$

$$M(A) = M(u,v) = \frac{1}{2} \int_a^b (uv' - vu') \, dx = \int_a^b uv' \, dx$$

will therefore satisfy the isoperimetric inequality

$$[L(\partial A)]^2 - 4\pi M(A) \geq 0.$$

This is, of course, the case for any simple closed smooth curve, whose interior is A.

One should also note that very wild sets A can be allowed. Indeed sets A that can be approximated by sets A_ν that satisfy the isoperimetric inequality and which are so that

$$L(\partial A_\nu) \to L(\partial A) \quad \text{and} \quad M(A_\nu) \to M(A), \text{ as } \nu \to \infty$$

also verify the inequality.

6.2.1 Exercises

Exercise 6.2.1 Prove Theorem 6.1 in an analogous manner as that of Example 2.24.

Exercise 6.2.2 Let

$$(P) \ \inf\left\{ I(u) = \int_{-1}^1 (u'^2 - \pi^2 u^2) \, dx : u \in X \right\} = m$$

where

$$X = \left\{ u \in W^{1,2}_{\text{per}}(-1,1) : \int_{-1}^1 u(x) \, dx = 0 \right\}.$$

Prove that, for any $\alpha, \beta \in \mathbb{R}$,

$$u(x) = \alpha \cos(\pi x) + \beta \sin(\pi x)$$

are the only minimizers in X.

Suggestion. We have seen in Theorem 6.1 that $m = 0$ and the minimum is attained in $X \cap C^2[-1,1]$ if and only if $u(x) = \alpha \cos(\pi x) + \beta \sin(\pi x)$, for any $\alpha, \beta \in \mathbb{R}$. Show that any minimizer of (P) is $C^2[-1,1]$. Conclude.

Exercise 6.2.3 Prove Step 1 of Theorem 6.4 for any $u, v \in C^1_{\text{per}}([a,b])$.

The following properties then hold.

  (i) $A + \overline{B_R} = \{x \in \mathbb{R}^n : d(x, A) \leq R\}$.

  (ii) If A is convex, then $A + B_R$ is also convex.

  Proof. (i) Let $x \in A + \overline{B_R}$ and

$$X = \{x \in \mathbb{R}^n : d(x, A) \leq R\}.$$

We then have that $x = a + b$ for some $a \in A$ and $b \in \overline{B_R}$, and hence

$$|x - a| = |b| \leq R$$

which implies that $x \in X$. Conversely, since A is compact, we can find, for every $x \in X$, an element $a \in A$ so that $|x - a| \leq R$. Letting $b = x - a$, we have indeed found that $x \in A + \overline{B_R}$.

  (ii) This is trivial. ∎

  We now examine the meaning of the proposition in a simple example.

Example 6.9 If A is a rectangle in $\mathbb{R}^2$, we find that $A + \overline{B_R}$ is given by the figure below. Anticipating, a little, on the following results we see that we have

$$M(A + \overline{B_R}) = M(A) + RL(\partial A) + R^2\pi$$

where $L(\partial A)$ is the perimeter of ∂A.

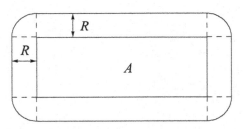

Figure 6.1: $A + B_R$

  We now define the meaning (cf. Berger [10], Section 12.10.9.3) of $L(\partial A)$ and $M(A)$.

Definition 6.10 *Let $n \geq 2$ and $A \subset \mathbb{R}^n$ be a compact set. We define*

  (i) $M(A)$ as the Lebesgue measure of A;

Remark 6.14 (i) The same proof establishes that the function $A \to (M(A))^{1/n}$ is concave. We thus have

$$[M(\lambda A + (1-\lambda)B)]^{1/n} \geq \lambda [M(A)]^{1/n} + (1-\lambda)[M(B)]^{1/n}$$

for every compact $A, B \subset \mathbb{R}^n$ and for every $\lambda \in [0,1]$.

(ii) One can even show that the function is strictly concave. This implies that the inequality in the theorem is strict unless A and B are homothetic.

Example 6.15 Let $n = 1$.

(i) If $A = [a,b]$, $B = [c,d]$, we have $A + B = [a+c, \ b+d]$ and

$$M(A+B) = M(A) + M(B).$$

(ii) If $A = [0,1]$, $B = [0,1] \cup [2,3]$, we find $A + B = [0,4]$ and hence

$$M(A+B) = 4 > M(A) + M(B) = 3.$$

We prove Theorem 6.13 at the end of the section. We are now in a position to state and prove the isoperimetric inequality.

Theorem 6.16 (Isoperimetric inequality) *Let $A \subset \mathbb{R}^n$, $n \geq 2$, be a compact set, $L = L(\partial A)$, $M = M(A)$ and ω_n be as above, then the following inequality holds*

$$L^n - n^n \omega_n M^{n-1} \geq 0.$$

Furthermore equality holds, among all convex sets, if and only if A is a ball.

Remark 6.17 (i) The proof that we give is also valid in the case $n = 2$. However it is unduly complicated and less precise than the one given in the preceding section.

(ii) Concerning the uniqueness that we do not prove below (cf. Berger [10], Section 12.11), we should point out that it is uniqueness only among convex sets. In dimension 2, we did not need this restriction; since for a non-convex set A, its convex hull has larger area and smaller perimeter. In higher dimensions this is not true anymore. In the case $n \geq 3$, one can still obtain uniqueness by assuming some regularity of the boundary ∂A, in order to avoid "hairy" spheres (i.e. sets that have zero n and $(n-1)$ measures but non-zero lower dimensional measures).

The theorem follows from (6.2) by setting $\lambda = 1/2$. If we let

$$A = \prod_{i=1}^{n} (a_i, b_i) \quad \text{and} \quad B = \prod_{i=1}^{n} (c_i, d_i)$$

we obtain

$$\lambda A + (1 - \lambda) B = \prod_{i=1}^{n} (\lambda a_i + (1 - \lambda) c_i, \lambda b_i + (1 - \lambda) d_i).$$

Setting, for $1 \leq i \leq n$,

$$u_i = \frac{b_i - a_i}{\lambda (b_i - a_i) + (1 - \lambda) (d_i - c_i)}, \quad v_i = \frac{d_i - c_i}{\lambda (b_i - a_i) + (1 - \lambda) (d_i - c_i)} \quad (6.3)$$

we find that

$$\lambda u_i + (1 - \lambda) v_i = 1, \ 1 \leq i \leq n, \tag{6.4}$$

$$\frac{M(A)}{M(\lambda A + (1 - \lambda) B)} = \prod_{i=1}^{n} u_i \quad \text{and} \quad \frac{M(B)}{M(\lambda A + (1 - \lambda) B)} = \prod_{i=1}^{n} v_i. \tag{6.5}$$

We now combine (6.1), (6.4) and (6.5) to deduce that

$$\frac{\lambda [M(A)]^{1/n} + (1 - \lambda) [M(B)]^{1/n}}{[M(\lambda A + (1 - \lambda) B)]^{1/n}} = \lambda \prod_{i=1}^{n} u_i^{1/n} + (1 - \lambda) \prod_{i=1}^{n} v_i^{1/n}$$

$$\leq \lambda \sum_{i=1}^{n} \frac{u_i}{n} + (1 - \lambda) \sum_{i=1}^{n} \frac{v_i}{n}$$

$$= \frac{1}{n} \sum_{i=1}^{n} (\lambda u_i + (1 - \lambda) v_i) = 1$$

and hence the result.

Step 3. We now prove (6.2) for any A and B of the form

$$A = \bigcup_{\mu=1}^{M} A_\mu, \ B = \bigcup_{\nu=1}^{N} B_\nu$$

where $A_\mu, B_\nu \in \mathcal{F}$, $A_\nu \cap A_\mu = B_\nu \cap B_\mu = \emptyset$ if $\mu \neq \nu$. The proof is achieved through induction on $M + N$. Step 2 has established the result when $M = N = 1$. We assume now that $M > 1$. We then choose $i \in \{1, \cdots, n\}$ and $a \in \mathbb{R}$ such that if

$$A^+ = A \cap \{x \in \mathbb{R}^n : x_i > a\} \quad \text{and} \quad A^- = A \cap \{x \in \mathbb{R}^n : x_i < a\}$$

are separated by $\{x : x_i = \lambda a + (1 - \lambda) b\}$ and thus

$$M\left(\lambda A + (1 - \lambda) B\right) = M\left(\lambda A^+ + (1 - \lambda) B^+\right) + M\left(\lambda A^- + (1 - \lambda) B^-\right).$$

Applying the hypothesis of induction to A^+, B^+ and A^-, B^-, we deduce that

$$M\left(\lambda A + (1 - \lambda) B\right) \geq \left[\lambda\left[M\left(A^+\right)\right]^{1/n} + (1 - \lambda)\left[M\left(B^+\right)\right]^{1/n}\right]^n$$
$$+ \left[\lambda\left[M\left(A^-\right)\right]^{1/n} + (1 - \lambda)\left[M\left(B^-\right)\right]^{1/n}\right]^n.$$

Using (6.7) we obtain

$$M\left(\lambda A + (1 - \lambda) B\right) \geq \frac{M\left(A^+\right)}{M\left(A\right)}\left[\lambda\left[M\left(A\right)\right]^{1/n} + (1 - \lambda)\left[M\left(B\right)\right]^{1/n}\right]^n$$
$$+ \frac{M\left(A^-\right)}{M\left(A\right)}\left[\lambda\left[M\left(A\right)\right]^{1/n} + (1 - \lambda)\left[M\left(B\right)\right]^{1/n}\right]^n.$$

The identity (6.6) and the above inequality imply then (6.2).

Step 4. We finally show (6.2) for any compact set, concluding thus the proof of the theorem. Let $\epsilon > 0$, we can then approximate the compact sets A and B, by A_ϵ and B_ϵ as in Step 3, so that

$$|M\left(A\right) - M\left(A_\epsilon\right)|, \quad |M\left(B\right) - M\left(B_\epsilon\right)| \leq \epsilon, \tag{6.8}$$

$$|M\left(\lambda A + (1 - \lambda) B\right) - M\left(\lambda A_\epsilon + (1 - \lambda) B_\epsilon\right)| \leq \epsilon. \tag{6.9}$$

Applying (6.2) to A_ϵ, B_ϵ, using (6.8) and (6.9), we obtain, after passing to the limit as $\epsilon \to 0$, the claim

$$\left[M\left(\lambda A + (1 - \lambda) B\right)\right]^{1/n} \geq \lambda\left[M\left(A\right)\right]^{1/n} + (1 - \lambda)\left[M\left(B\right)\right]^{1/n}.$$

This achieves the proof of the theorem. ■

6.3.1   Exercises

Exercise 6.3.1 Let $A, B \subset \mathbb{R}$ be compact,

$$\overline{a} = \min\{a : a \in A\} \quad \text{and} \quad \overline{b} = \max\{b : b \in B\}.$$

Prove that

$$(\overline{a} + B) \cup (\overline{b} + A) \subset A + B$$

and deduce that

$$M\left(A\right) + M\left(B\right) \leq M\left(A + B\right).$$

We therefore obtain that, for $0 < \alpha < \lambda \leq 1$,

$$|u_\lambda(x) - u_\lambda(y)| \leq \frac{2}{2^{\lambda-\alpha} - 1} |x - y|^\alpha.$$

Note that the function u_λ, which is essentially the classical Weierstrass example, is nowhere differentiable for any $\lambda \in (0, 1]$ (see for example [95]). ∎

Exercise 1.2.2. (i) We have

$$\|uv\|_{C^{0,\alpha}} = \|uv\|_{C^0} + [uv]_{C^{0,\alpha}}.$$

Since

$$
\begin{aligned}
[uv]_{C^{0,\alpha}} &\leq \sup \frac{|u(x)v(x) - u(y)v(y)|}{|x - y|^\alpha} \\
&\leq \|u\|_{C^0} \sup \frac{|v(x) - v(y)|}{|x - y|^\alpha} + \|v\|_{C^0} \sup \frac{|u(x) - u(y)|}{|x - y|^\alpha}
\end{aligned}
$$

we deduce that

$$
\begin{aligned}
\|uv\|_{C^{0,\alpha}} &\leq \|u\|_{C^0} \|v\|_{C^0} + \|u\|_{C^0} [v]_{C^{0,\alpha}} + \|v\|_{C^0} [u]_{C^{0,\alpha}} \\
&\leq 2 \|u\|_{C^{0,\alpha}} \|v\|_{C^{0,\alpha}}.
\end{aligned}
$$

(ii) The inclusion $C^{0,\alpha} \subset C^0$ is obvious. Let us show that $C^{0,\beta} \subset C^{0,\alpha}$. Observe that

$$\sup_{\substack{x,y \in \overline{\Omega} \\ 0 < |x-y| < 1}} \left\{ \frac{|u(x) - u(y)|}{|x - y|^\alpha} \right\} \leq \sup_{\substack{x,y \in \overline{\Omega} \\ 0 < |x-y| < 1}} \left\{ \frac{|u(x) - u(y)|}{|x - y|^\beta} \right\} \leq [u]_{C^{0,\beta}}.$$

Since

$$\sup_{\substack{x,y \in \overline{\Omega} \\ |x-y| \geq 1}} \left\{ \frac{|u(x) - u(y)|}{|x - y|^\alpha} \right\} \leq \sup_{x,y \in \overline{\Omega}} \{|u(x)| - u(y)\} \leq 2 \|u\|_{C^0}$$

we get

$$
\begin{aligned}
\|u\|_{C^{0,\alpha}} &= \|u\|_{C^0} + [u]_{C^{0,\alpha}} \\
&\leq \|u\|_{C^0} + \max\{2 \|u\|_{C^0}, [u]_{C^{0,\beta}}\} \leq 3 \|u\|_{C^{0,\beta}}.
\end{aligned}
$$

(iii) We now assume that Ω is bounded and convex, and let us show that $C^1(\overline{\Omega}) \subset C^{0,1}(\overline{\Omega})$. Let $x, y \in \overline{\Omega}$. Since $\overline{\Omega}$ is convex, we have that $[x, y] \subset \overline{\Omega}$ (where $[x, y]$ denotes the segment joining x to y). We can therefore write

$$u(x) - u(y) = \int_0^1 \frac{d}{dt} u(y + t(x - y)) \, dt = \int_0^1 \langle \nabla u(y + t(x - y)); x - y \rangle \, dt$$

(ii) We now discuss the case $k = 1$; for higher derivatives $(k \geq 2)$ we proceed with a straightforward induction. Let $u \in C^{1,\alpha}(\overline{\Omega}; \mathbb{R}^n)$, $u = u(x) = (u^1, \cdots, u^n)$, and $v \in C^{1,\alpha}(u(\overline{\Omega}))$, $v = v(y)$. Let $i = 1, \cdots, n$ and consider

$$(v \circ u)_{x_i} = \sum_{j=1}^{n} (v_{y_j} \circ u) u_{x_i}^j.$$

Note that it follows from Proposition 1.9 (iii) that $u \in C^{0,1}(\overline{\Omega}; \mathbb{R}^n)$ and hence, by (i) above, that $v_{y_j} \circ u \in C^{0,\alpha}(\overline{\Omega})$. Appealing to Proposition 1.9 (i), we deduce that

$$\left(v_{y_j} \circ u\right) u_{x_i}^j \in C^{0,\alpha}(\overline{\Omega})$$

which in turn implies $v \circ u \in C^{1,\alpha}(\overline{\Omega})$. $\blacksquare$

Exercice 1.2.5. **(i)** We discuss the case of u_+, the other one being handled similarly.

1) Let us first check that indeed u_+ is an extension of u. Let $x \in \overline{\Omega}$, we therefore get

$$u(x) \leq u(y) + \gamma |x - y|^\alpha \text{ for every } y \in \overline{\Omega} \Rightarrow u(x) \leq u_+(x).$$

Now, clearly, choosing $y = x$ in the definition of u_+, leads to $u_+(x) \leq u(x)$. Thus u_+ is indeed an extension of u.

2) Let $x, z \in \mathbb{R}^n$. Assume, without loss of generality, that $u_+(z) \leq u_+(x)$. For every $\epsilon > 0$, we can find $y_z \in \overline{\Omega}$ such that

$$-\epsilon + u(y_z) + \gamma |z - y_z|^\alpha \leq u_+(z) \leq u(y_z) + \gamma |z - y_z|^\alpha.$$

We hence obtain

$$
\begin{aligned}
|u_+(x) - u_+(z)| &= u_+(x) - u_+(z) \\
&\leq u(y_z) + \gamma |x - y_z|^\alpha + \epsilon - u(y_z) - \gamma |z - y_z|^\alpha \\
&\leq \epsilon + \gamma |x - z|^\alpha.
\end{aligned}
$$

Letting $\epsilon \to 0$ we have the claim.

(ii) Let v be such that $[v]_{C^{0,\alpha}(\mathbb{R}^n)} = \gamma$. We therefore have for $x \in \mathbb{R}^n$ and for every $y \in \overline{\Omega}$ (and thus $v(y) = u(y)$)

$$-\gamma |x - y|^\alpha \leq v(x) - v(y) = v(x) - u(y) \leq \gamma |x - y|^\alpha.$$

This leads to

$$u(y) - \gamma |x - y|^\alpha \leq v(x) \leq u(y) + \gamma |x - y|^\alpha$$

and hence $u_-(x) \leq v(x) \leq u_+(x)$ as wished. $\blacksquare$

(iii) The inclusion $L^\infty(\Omega) \subset L^p(\Omega)$ is trivial. The other inclusions follow from Hölder inequality. Indeed we have

$$\int_\Omega |u|^q = \int_\Omega (|u|^q \cdot 1) \le \left(\int_\Omega |u|^{q \cdot p/q}\right)^{q/p} \left(\int_\Omega 1^{p/(p-q)}\right)^{(p-q)/p}$$

$$\le (\text{meas }\Omega)^{(p-q)/p} \left(\int_\Omega |u|^p\right)^{q/p}$$

and hence

$$\|u\|_{L^q} \le (\text{meas }\Omega)^{(p-q)/pq} \|u\|_{L^p}$$

which gives the desired inclusion.

If, however, the measure is not finite, the result is not valid as the simple example $\Omega = (1,\infty)$, $u(x) = 1/x$ shows; indeed we have $u \in L^2$ but $u \notin L^1$. $\blacksquare$

Exercise 1.3.2. A direct computation leads to

$$\|u_\nu\|_{L^p}^p = \int_0^1 |u_\nu(x)|^p \, dx = \int_0^{1/\nu} \nu^{\alpha p} dx = \nu^{\alpha p - 1}.$$

We therefore have that $u_\nu \to 0$ in L^p provided $\alpha p - 1 < 0$. If $\alpha = 1/p$, let us show that $u_\nu \rightharpoonup 0$ in L^p. We have to prove that for every $\varphi \in L^{p'}(0,1)$, the following convergence holds

$$\lim_{\nu \to \infty} \int_0^1 u_\nu(x) \varphi(x) \, dx = 0.$$

By a density argument, and since $\|u_\nu\|_{L^p} = 1$, it is sufficient to prove the result when φ is a step function, which means that there exist $0 = a_0 < a_1 < \cdots < a_I = 1$ so that $\varphi(x) = \alpha_i$ whenever $x \in (a_{i-1}, a_i)$, $1 \le i \le I$. We hence find, for ν sufficiently large, that

$$\int_0^1 u_\nu(x) \varphi(x) \, dx = \alpha_1 \int_{a_0}^{1/\nu} \nu^{1/p} dx = \alpha_1 \nu^{(1/p)-1} \to 0. \quad \blacksquare$$

Exercise 1.3.3. (i) We have to show that for every $\varphi \in L^\infty$, then

$$\lim_{\nu \to \infty} \int_\Omega (u_\nu v_\nu - uv) \varphi = 0.$$

Rewriting the integral we have

$$\int_\Omega (u_\nu v_\nu - uv) \varphi = \int_\Omega u_\nu (v_\nu - v) \varphi + \int_\Omega (u_\nu - u) v\varphi.$$

Hölder inequality leads to

$$|u_\nu(x)| \leq \left(\int_{-\infty}^{+\infty} \varphi(z)\,dz\right)^{1/p'} \left(\int_{-\infty}^{+\infty} \varphi(z)\left|u\left(x - \frac{z}{\nu}\right)\right|^p dz\right)^{1/p}.$$

Since $\int \varphi = 1$, we have, after interchanging the order of integration,

$$\|u_\nu\|_{L^p}^p = \int_{-\infty}^{+\infty} |u_\nu(x)|^p\,dx \leq \int_{-\infty}^{+\infty}\int_{-\infty}^{+\infty}\left\{\varphi(z)\left|u\left(x - \frac{z}{\nu}\right)\right|^p dz\right\}dx$$

$$\leq \int_{-\infty}^{+\infty}\left\{\varphi(z)\int_{-\infty}^{+\infty}\left|u\left(x - \frac{z}{\nu}\right)\right|^p dx\right\}dz \leq \|u\|_{L^p}^p.$$

(ii) The result follows, since φ is C^∞ and

$$u_\nu'(x) = \int_{-\infty}^{+\infty} \varphi_\nu'(x - y)\,u(y)\,dy.$$

(iii) Let $K \subset \mathbb{R}$ be a fixed compact. Since u is continuous, we have that for every $\epsilon > 0$, there exists $\delta = \delta(\epsilon, K) > 0$ so that

$$|y| \leq \delta \Rightarrow |u(x - y) - u(x)| \leq \epsilon, \ \forall x \in K.$$

Since $\varphi = 0$ if $|x| > 1$, $\int \varphi = 1$, and hence $\int \varphi_\nu = 1$, we find that

$$u_\nu(x) - u(x) = \int_{-\infty}^{+\infty} [u(x - y) - u(x)]\,\varphi_\nu(y)\,dy$$

$$= \int_{-1/\nu}^{1/\nu} [u(x - y) - u(x)]\,\varphi_\nu(y)\,dy.$$

Taking $x \in K$ and $\nu > 1/\delta$, we deduce that $|u_\nu(x) - u(x)| \leq \epsilon$, and thus the claim.

(iv) Since $u \in L^p(\mathbb{R})$ and $1 \leq p < \infty$, we deduce (see Theorem 1.13 (vi)) that for every $\epsilon > 0$, there exists $\overline{u} \in C_0(\mathbb{R})$ so that

$$\|u - \overline{u}\|_{L^p} \leq \epsilon. \tag{7.1}$$

Define then

$$\overline{u}_\nu(x) = (\varphi_\nu * \overline{u})(x) = \int_{-\infty}^{+\infty} \varphi_\nu(x - y)\,\overline{u}(y)\,dy.$$

Since $u - \overline{u} \in L^p$, it follows from (i) that

$$\|u_\nu - \overline{u}_\nu\|_{L^p} \leq \|u - \overline{u}\|_{L^p} \leq \epsilon. \tag{7.2}$$

Using Hölder inequality, (1.1), (7.3) and (7.6), we obtain that

$$\left| \int_0^1 u_\nu(x) \varphi(x) \, dx \right| \leq \epsilon \|u\|_{L^p} + \|v_\nu\|_{L^\infty} \|\psi'\|_{L^1} \leq \epsilon \|u\|_{L^p} + \frac{1}{\nu} \|u\|_{L^p} \|\psi'\|_{L^1} \, .$$

Let $\nu \to \infty$, we hence deduce that

$$0 \leq \limsup_{\nu \to \infty} \left| \int_0^1 u_\nu \varphi \, dx \right| \leq \epsilon \|u\|_{L^p} \, .$$

Since ϵ is arbitrary, we immediately have (7.5) and thus the result. ∎

Exercise 1.3.6. (i) Let $f \in C_0^\infty(\Omega)$ with $\int_\Omega f(x) \, dx = 1$, be a fixed function. Let $w \in C_0^\infty(\Omega)$ be arbitrary and

$$\psi(x) = w(x) - \left[\int_\Omega w(y) \, dy \right] f(x) \, .$$

We therefore have $\psi \in C_0^\infty(\Omega)$ and $\int \psi = 0$ which leads to

$$0 = \int_\Omega u(x) \psi(x) \, dx = \int_\Omega u(x) w(x) \, dx - \int_\Omega f(x) u(x) \, dx \cdot \int_\Omega w(y) \, dy$$

$$= \int_\Omega \left[u(x) - \int_\Omega u(y) f(y) \, dy \right] w(x) \, dx.$$

Appealing to Theorem 1.24, we deduce that $u(x) = \int u(y) f(y) \, dy = \text{constant}$ a.e.

   (ii) Let $\psi \in C_0^\infty(a, b)$, with $\int_a^b \psi = 0$, be arbitrary and define

$$\varphi(x) = \int_a^x \psi(t) \, dt \, .$$

Note that $\psi = \varphi'$ and $\varphi \in C_0^\infty(a, b)$. We may thus apply (i) and get the result. ∎

Exercise 1.3.7. We define $(N + 1)$ linear functionals on $C_0^\infty(\Omega)$ by

$$\Lambda(\psi) = \int_\Omega u(x) \psi(x) \, dx \quad \text{and} \quad \Lambda_i(\psi) = \int_\Omega \alpha_i(x) \psi(x) \, dx, \ i = 1, \cdots, N.$$

Note that our hypothesis guarantees that

$$\Lambda(\psi) = 0, \text{ for every } \psi \in C_0^\infty(\Omega) \text{ with } \Lambda_i(\psi) = 0, \ i = 1, \cdots, N.$$

Lemma 3.9 page 62 in [93] implies then that there exist constants $a_1, \cdots, a_N \in \mathbb{R}$ such that

$$\Lambda = \sum_{i=1}^N a_i \Lambda_i \, .$$

7.1.3 Sobolev spaces

Exercise 1.4.1. Let $\sigma_{n-1} = \text{meas}\,(\partial B_1\,(0))$ (i.e. $\sigma_1 = 2\pi$, $\sigma_2 = 4\pi$, $\cdots$).
  (i) The result follows from the following observation

$$\|u\|_{L^p}^p = \int_{B_R} |u\,(x)|^p\,dx = \sigma_{n-1} \int_0^R r^{n-1}\,|f\,(r)|^p\,dr.$$

(ii) We find, if $x \neq 0$, that

$$u_{x_i} = f'\,(|x|)\,\frac{x_i}{|x|} \;\Rightarrow\; |\nabla u\,(x)| = |f'\,(|x|)|\,.$$

Assume, for a moment, that we already proved that u is weakly differentiable in B_R, then

$$\|\nabla u\|_{L^p}^p = \sigma_{n-1} \int_0^R r^{n-1}\,|f'\,(r)|^p\,dr,$$

which is the claim.

Let us now show that u_{x_i}, as above, is indeed the weak derivative (with respect to x_i) of u. We have to prove that, for every $\varphi \in C_0^\infty\,(B_R)$,

$$\int_{B_R} u\varphi_{x_i}\,dx = -\int_{B_R} \varphi u_{x_i}\,dx. \tag{7.7}$$

Let $\epsilon > 0$ be sufficiently small and observe that (recall that $\varphi = 0$ on ∂B_R)

$$
\begin{aligned}
\int_{B_R} u\varphi_{x_i}\,dx &= \int_{B_R \setminus B_\epsilon} u\varphi_{x_i}\,dx + \int_{B_\epsilon} u\varphi_{x_i}\,dx \\
&= -\int_{B_R \setminus B_\epsilon} \varphi u_{x_i}\,dx - \int_{\partial B_\epsilon} u\varphi\,\frac{x_i}{|x|}\,d\sigma + \int_{B_\epsilon} u\varphi_{x_i}\,dx \\
&= -\int_{B_R} \varphi u_{x_i}\,dx + \int_{B_\epsilon} \varphi u_{x_i}\,dx + \int_{B_\epsilon} u\varphi_{x_i}\,dx - \int_{\partial B_\epsilon} u\varphi\,\frac{x_i}{|x|}\,d\sigma.
\end{aligned}
$$

Since the elements φu_{x_i} and $u\varphi_{x_i}$ are both in $L^1\,(B_R)$, we deduce (see Exercise 1.3.9) that

$$\lim_{\epsilon \to 0} \int_{B_\epsilon} \varphi u_{x_i}\,dx = \lim_{\epsilon \to 0} \int_{B_\epsilon} u\varphi_{x_i}\,dx = 0.$$

Moreover, by hypothesis, we have the claim (i.e. (7.7)), since

$$\left| \int_{\partial B_\epsilon} u\varphi\,\frac{x_i}{|x|}\,d\sigma \right| \leq \sigma_{n-1}\,\|\varphi\|_{L^\infty}\,\epsilon^{n-1}\,|f\,(\epsilon)| \to 0, \quad \text{as } \epsilon \to 0.$$

(iii) 1) The first example follows at once and gives

$$\psi \in L^p \;\Leftrightarrow\; sp < n \quad \text{and} \quad \psi \in W^{1,p} \;\Leftrightarrow\; (s+1)\,p < n.$$

and by the properties of Lebesgue integrals (see Exercise 1.3.9) the quantity $\left(\int_y^x |u'(t)|^p \, dt \right)^{1/p}$ tends to 0 as $|x - y|$ tends to 0. ∎

Exercise 1.4.4. Observe first that if $v \in W^{1,p}(a,b)$, $p > 1$ and $y < x$, then

$$|v(x) - v(y)| = \left| \int_y^x v'(z) \, dz \right|$$

$$\leq \left(\int_y^x |v'(z)|^p \, dz \right)^{1/p} \left(\int_y^x dz \right)^{1/p'} \leq \|v'\|_{L^p} |x - y|^{1/p'}. \tag{7.8}$$

Let us now show that if $u_\nu \rightharpoonup u$ in $W^{1,p}$, then $u_\nu \to u$ in L^∞. Without loss of generality, we can take $u \equiv 0$. Assume, for the sake of contradiction, that $u_\nu \nrightarrow 0$ in L^∞. We can therefore find $\epsilon > 0$, $\{\nu_i\}$ so that

$$\|u_{\nu_i}\|_{L^\infty} \geq \epsilon, \; \nu_i \to \infty. \tag{7.9}$$

From (7.8) we have that the subsequence $\{u_{\nu_i}\}$ is equicontinuous (note also that by Theorem 1.43 and Theorem 1.20 (iii) we have $\|u_{\nu_i}\|_{L^\infty} \leq c' \|u_{\nu_i}\|_{W^{1,p}} \leq c$) and hence from Ascoli-Arzelà theorem, we find, up to a subsequence,

$$u_{\nu_{i_j}} \to v \text{ in } L^\infty. \tag{7.10}$$

We, however, must have $v = 0$ since (7.10) implies $u_{\nu_{i_j}} \rightharpoonup v$ in L^p and by uniqueness of the limits (we already know that $u_{\nu_{i_j}} \rightharpoonup u = 0$ in L^p) we deduce that $v = 0$ a.e., which contradicts (7.9). ∎

Exercise 1.4.5. From Theorem 1.20 we have that there exist $u, v_k \in L^p(\Omega)$, $k = 1, \cdots, n$, and a subsequence such that

$$u_{\nu_i} \rightharpoonup u \text{ in } L^p \quad \text{and} \quad (u_{\nu_i})_{x_k} \rightharpoonup v_k \text{ in } L^p.$$

Moreover $v_k = u_{x_k}$, since for every $\varphi \in C_0^\infty(\Omega)$

$$\int v_k \varphi = \lim_{\nu_i \to \infty} \int (u_{\nu_i})_{x_k} \varphi = -\lim_{\nu_i \to \infty} \int u_{\nu_i} \varphi_{x_k} = -\int u \varphi_{x_k}. \; ∎$$

Exercise 1.4.6. It is enough to prove the result when $u = 0$. Let $\varphi \in L^{p'}(\Omega)$ $(1 \leq p' < \infty)$ and $\epsilon > 0$. We can find $\psi \in C_0^\infty(\Omega)$ such that

$$\|\varphi - \psi\|_{L^{p'}} \leq \epsilon.$$

Note that, since $u_\nu \in W^{1,p}$ and $\psi \in C_0^\infty(\Omega)$, we can write, for every $i = 1, \cdots, n$,

$$\int (u_\nu)_{x_i} \varphi = \int (u_\nu)_{x_i} \psi + \int (u_\nu)_{x_i} (\varphi - \psi)$$

$$= -\int u_\nu \psi_{x_i} + \int (u_\nu)_{x_i} (\varphi - \psi).$$

Exercice 1.4.9. **(i)** Let $\epsilon > 0$ and $u \in W_0^{1,p}(\Omega)$. By definition, we can find $u_\epsilon \in C_0^\infty(\Omega)$ such that

$$\|u\|_{L^p} \leq \|u_\epsilon\|_{L^p} + \epsilon \quad \text{and} \quad \|\nabla u_\epsilon\|_{L^p} \leq \|\nabla u\|_{L^p} + \epsilon.$$

Applying Poincaré inequality for C_0^∞ functions, we can find $\gamma = \gamma(\Omega, p) > 0$ such that

$$\|u\|_{L^p} \leq \|u_\epsilon\|_{L^p} + \epsilon \leq \gamma \|\nabla u_\epsilon\|_{L^p} + \epsilon \leq \gamma \|\nabla u\|_{L^p} + (\gamma + 1)\epsilon.$$

Since ϵ is arbitrary, we have the claim.

(ii) We have

$$u(x_1, x_2, \cdots, x_n) = u(-R, x_2, \cdots, x_n) + \int_{-R}^{x_1} \frac{d}{dt} u(t, x_2, \cdots, x_n) \, dt$$

and thus

$$|u(x)| \leq \int_{-R}^{x_1} |u_{x_1}(t, x_2, \cdots, x_n)| \, dt \leq \int_{-R}^{R} |u_{x_1}(t, x_2, \cdots, x_n)| \, dt.$$

Appealing to Jensen inequality (cf. Theorem 1.52), we find

$$|u(x)|^p \leq (2R)^p \left(\frac{1}{2R} \int_{-R}^{R} |u_{x_1}(t, x_2, \cdots, x_n)| \, dt \right)^p$$

$$\leq (2R)^{p-1} \int_{-R}^{R} |u_{x_1}(t, x_2, \cdots, x_n)|^p \, dt.$$

We hence get, integrating with respect to $x_2, \cdots, x_n$,

$$\int_{-R}^{R} \cdots \int_{-R}^{R} |u(x_1, x_2, \cdots, x_n)|^p \, dx_2 \cdots dx_n$$

$$\leq (2R)^{p-1} \int_Q |u_{x_1}(t, x_2, \cdots, x_n)|^p \, dt dx_2 \cdots dx_n.$$

Integrating once more, this time with respect to x_1, we have

$$\|u\|_{L^p(Q)}^p \leq (2R)^p \|\nabla u\|_{L^p(Q)}^p.$$

(iii) We first choose $R > 0$ sufficiently large so that $\Omega \subset \overline{\Omega} \subset Q = (-R, R)^n$. Let $u \in C_0^\infty(\Omega)$ be extended by 0 outside Ω. Apply (ii) to find

$$\|u\|_{L^p(\Omega)} = \|u\|_{L^p(Q)} \leq (2R) \|\nabla u\|_{L^p(Q)} = (2R) \|\nabla u\|_{L^p(\Omega)}.$$

1) We start by observing that, by Hölder inequality, we have, for every $0 < \epsilon < 1/2$,

$$\int_{B_{2\epsilon}} |u| \, dx \leq \left(\int_{B_{2\epsilon}} |u|^p \, dx \right)^{1/p} (\text{meas } B_{2\epsilon})^{1/p'} \leq \|u\|_{L^p} (\text{meas } B_{2\epsilon})^{1/p'}$$

and thus there exists a constant $\gamma_1 = \gamma_1 (n, \|u\|_{L^p}) > 0$, such that

$$\int_{B_{2\epsilon}} |u| \, dx \leq \gamma_1 \epsilon^{n(p-1)/p}.$$

2) We next define, for $0 < \epsilon < 1/2$, a function $\rho_\epsilon \in C^\infty ([-1,1])$, $0 \leq \rho_\epsilon \leq 1$,

$$\rho_\epsilon (t) = \begin{cases} 1 & \text{if } |t| \leq \epsilon \\ 0 & \text{if } |t| \geq 2\epsilon \end{cases} \quad \text{and} \quad |\rho'_\epsilon (t)| \leq \frac{\gamma_2}{\epsilon}$$

for a certain $\gamma_2 > 0$. We then set

$$u_\epsilon (x) = (1 - \rho_\epsilon (|x|)) u (x).$$

Observe that $u_\epsilon \in C^1 (\overline{B_1})$ and, since $u_{x_i} = 0$ in $\overline{B_1} \smallsetminus \{0\}$,

$$\frac{\partial u_\epsilon}{\partial x_i} (x) = \begin{cases} -\rho'_\epsilon (|x|) \frac{x_i}{|x|} u (x) & \text{if } \epsilon \leq |x| \leq 2\epsilon \\ 0 & \text{if } |x| \geq 2\epsilon \text{ or } |x| \leq \epsilon. \end{cases}$$

We moreover obtain, for every $\varphi \in C_0^\infty (B_1)$,

$$\left| \int_{B_1} u_\epsilon \varphi_{x_i} dx \right| = \left| \int_{B_1} \frac{\partial u_\epsilon}{\partial x_i} \varphi dx \right| \leq \frac{\gamma_2}{\epsilon} \|\varphi\|_{L^\infty} \int_{B_{2\epsilon}} |u| \, dx.$$

3) Combining together the two observations, we get

$$\left| \int_{B_1} u \varphi_{x_i} dx \right| = \left| \int_{B_1} u_\epsilon \varphi_{x_i} dx + \int_{B_1} \rho_\epsilon u \varphi_{x_i} dx \right|$$

$$= \left| \int_{B_1} u_\epsilon \varphi_{x_i} dx + \int_{B_{2\epsilon}} \rho_\epsilon u \varphi_{x_i} dx \right|$$

$$\leq \frac{\gamma_2}{\epsilon} \|\varphi\|_{L^\infty} \int_{B_{2\epsilon}} |u| \, dx + \|\varphi\|_{W^{1,\infty}} \int_{B_{2\epsilon}} |u| \, dx$$

$$\leq \gamma_1 \epsilon^{n(p-1)/p} \|\varphi\|_{W^{1,\infty}} \left(\frac{\gamma_2}{\epsilon} + 1 \right).$$

The result follows by letting $\epsilon \to 0$, since $p > n/(n-1)$ is equivalent to $n(p-1)/p > 1$. ∎

Returning to the definition of φ and y, we find, as wished,

$$f(y) \geq f(x) + \langle \nabla f(x) ; y - x \rangle.$$

(iii) $\Rightarrow$ **(iv)** Choose $y = x + \epsilon v$ with $\epsilon \neq 0$ and apply (iii) to get

$$\left\langle \frac{\nabla f(x + \epsilon v) - \nabla f(x)}{\epsilon} ; v \right\rangle \geq 0.$$

This implies, letting $\epsilon \to 0$,

$$\langle \nabla^2 f(x) v ; v \rangle \geq 0.$$

(iv) $\Rightarrow$ **(iii)** Write

$$\langle \nabla f(x) - \nabla f(y) ; x - y \rangle = \int_0^1 \left\langle \frac{d}{dt} \nabla f(y + t(x - y)) ; x - y \right\rangle dt$$

$$= \int_0^1 \langle \nabla^2 f(y + t(x - y))(x - y) ; x - y \rangle dt$$

and apply (iv) to have the claim. ∎

Exercise 1.5.2. Since f is convex, we have, for every $\alpha, \beta \in \mathbb{R}$,

$$f(\alpha) \geq f(\beta) + f'(\beta)(\alpha - \beta).$$

Choose then $\alpha = u(x)$ and $\beta = (1/\operatorname{meas} \Omega) \int_\Omega u(x)\, dx$, and integrate to get the inequality. ∎

Exercise 1.5.3. We first observe that f^* is even and therefore we consider only $x^* \geq 0$. We easily find that if $x^* > 1$, then $f^*(x^*) = +\infty$. When $x^* = 1$, we get, in a straightforward way, that $f^*(x^*) = 0$. We now discuss the case $0 \leq x^* < 1$. It is clear, in this case, that the supremum in the definition of f^* is attained at the point

$$x^* = \frac{x}{\sqrt{1 + x^2}} \Leftrightarrow x = \frac{x^*}{\sqrt{1 - x^{*2}}}.$$

We thus get

$$f^*(x^*) = xx^* - \sqrt{1 + x^2} = -\sqrt{1 - x^{*2}}.$$

We have therefore found that

$$f^*(x^*) = \begin{cases} -\sqrt{1 - x^{*2}} & \text{if } |x^*| \leq 1 \\ +\infty & \text{otherwise.} \end{cases}$$

Note, in passing, that $f(x) = \sqrt{1 + x^2}$ is strictly convex over $\mathbb{R}$. ∎

and therefore

$$f^{**}(X) = \sup_{X^* \in \mathbb{R}^{2 \times 2}} \{\langle X; X^* \rangle - f^*(X^*)\} \equiv 0. \quad \blacksquare$$

Exercise 1.5.6. (i) Let $x^*, y^* \in \mathbb{R}^n$ and $\lambda \in [0,1]$. It follows from the definition that

$$\begin{aligned}
f^*(\lambda x^* + (1-\lambda) y^*) &= \sup_{x \in \mathbb{R}^n} \{\langle x; \lambda x^* + (1-\lambda) y^* \rangle - f(x)\} \\
&= \sup_x \{\lambda (\langle x; x^* \rangle - f(x)) + (1-\lambda)(\langle x; y^* \rangle - f(x))\} \\
&\leq \lambda \sup_x \{\langle x; x^* \rangle - f(x)\} + (1-\lambda) \sup_x \{\langle x; y^* \rangle - f(x)\} \\
&\leq \lambda f^*(x^*) + (1-\lambda) f^*(y^*).
\end{aligned}$$

(ii) For this part we can refer to Theorem I.10 in Brézis [14], Theorem 2.43 in [31, 2nd edition] or Theorem 12.2 (coupled with Corollaries 10.1.1 and 12.1.1) in Rockafellar [91].

(iii) Since $f^{**} \leq f$, we find that $f^{***} \geq f^*$. Furthermore, by definition of f^{**}, we find, for every $x \in \mathbb{R}^n$, $x^* \in \mathbb{R}^n$,

$$\langle x; x^* \rangle - f^{**}(x) \leq f^*(x^*).$$

Taking the supremum over all x in the left-hand side of the inequality, we get $f^{***} \leq f^*$, and hence the claim.

(iv) By definition of f^*, we have

$$f^*(\nabla f(x)) = \sup_y \{\langle y; \nabla f(x) \rangle - f(y)\} \geq \langle x; \nabla f(x) \rangle - f(x)$$

and hence

$$f(x) + f^*(\nabla f(x)) \geq \langle x; \nabla f(x) \rangle.$$

We next show the reverse inequality. Since f is convex, we have

$$f(y) \geq f(x) + \langle y - x; \nabla f(x) \rangle,$$

which means that

$$\langle x; \nabla f(x) \rangle - f(x) \geq \langle y; \nabla f(x) \rangle - f(y).$$

Taking the supremum over all y, we have indeed obtained the reverse inequality and thus the proof is complete.

(v) We refer to the bibliography, in particular to Mawhin-Willem [76] page 35, Rockafellar [91] (Theorems 23.5, 26.3 and 26.5 as well as Corollary 25.5.1) and

Exercise 1.5.8. (i) Let $h > 0$ and let $\{e_1, \cdots, e_n\}$ be the Euclidean basis. Use the convexity of f (in the form of (ii) of Theorem 1.51) and (1.15) to write

$$h \frac{\partial f}{\partial x_i}(x) \leq f(x + he_i) - f(x) \leq \alpha_1 (1 + |x + he_i|^p) + \alpha_1 (1 + |x|^p)$$

$$-h \frac{\partial f}{\partial x_i}(x) \leq f(x - he_i) - f(x) \leq \alpha_1 (1 + |x - he_i|^p) + \alpha_1 (1 + |x|^p).$$

We can therefore find $\widetilde{\alpha}_1 > 0$, so that

$$\left| \frac{\partial f}{\partial x_i}(x) \right| \leq \widetilde{\alpha}_1 \frac{(1 + |x|^p + |h|^p)}{h}.$$

Choosing $h = 1 + |x|$, we can surely find $\alpha_2 > 0$ so that (1.16) is satisfied, i.e.

$$\left| \frac{\partial f}{\partial x_i}(x) \right| \leq \alpha_2 \left(1 + |x|^{p-1} \right), \ \forall x \in \mathbb{R}.$$

The inequality (1.17) is then a consequence of (1.16), since

$$f(x) - f(y) = \int_0^1 \frac{d}{ds} f(y + s(x - y)) \, ds = \int_0^1 \langle \nabla f(y + s(x - y)); x - y \rangle \, ds.$$

(ii) Note that the convexity of f is essential in the above argument. Indeed, taking, for example $n = 1$ and $f(x) = x + \sin x^2$, we find that f satisfies (1.15) with $p = 1$, but it does not verify (1.16). More sophisticated examples show that the result does not carry either to the second derivative of a convex function f.

(iii) Of course if $\partial f / \partial x_i$ satisfies (1.16), we have by straight integration

$$f(x) = f(0) + \int_0^1 \frac{d}{ds} f(sx) \, ds = f(0) + \int_0^1 \langle \nabla f(sx); x \rangle \, ds$$

that f verifies (1.15), even if f is not convex. ∎

Exercise 1.5.9. With an easy argument, see [31] for details, the proof below in fact gives that f is locally Lipschitz continuous. We divide the proof into two steps.

Step 1. Let $x_0 \in \mathbb{R}$, $\{e_1, \cdots, e_n\}$ be the Euclidean basis and

$$a > \max_{i=1,\cdots,n} \{|f(x_0 + e_i) - f(x_0)|, |f(x_0 - e_i) - f(x_0)|\}.$$

It is then easy to see that if we denote by

$$|x|_1 = \sum_{i=1}^n |x_i|$$

7.2   Chapter 2. Classical methods

7.2.1   Euler-Lagrange equation

Exercise 2.2.1. The proof is almost identical to that of the theorem. The Euler-Lagrange equation becomes then a system of ordinary differential equations, namely, if $u = (u^1, \cdots, u^N)$ and $\xi = (\xi^1, \cdots, \xi^N)$, we have

$$\frac{d}{dx} \left[f_{\xi^i} (x, \overline{u}, \overline{u}') \right] = f_{u^i} (x, \overline{u}, \overline{u}'), \ i = 1, \cdots, N. \quad \blacksquare$$

Exercise 2.2.2. We proceed as in the theorem. We let

$$X = \left\{ u \in C^n ([a, b]) : u^{(j)} (a) = \alpha_j, \ u^{(j)} (b) = \beta_j, \ 0 \le j \le n - 1 \right\}$$

If $\overline{u} \in X \cap C^{2n} ([a, b])$ is a minimizer of (P) we have $I (\overline{u} + \epsilon v) \ge I (\overline{u}), \ \forall \epsilon \in \mathbb{R}$ and $\forall v \in C_0^\infty (a, b)$. Letting $f = f (x, u, \xi_1, \cdots, \xi_n)$ and using the fact that

$$\frac{d}{d\epsilon} I (\overline{u} + \epsilon v)|_{\epsilon = 0} = 0$$

we find

$$\int_a^b \left\{ f_u \left(x, \overline{u}, \cdots, \overline{u}^{(n)} \right) v + \sum_{i=1}^n f_{\xi_i} \left(x, \overline{u}, \cdots, \overline{u}^{(n)} \right) v^{(i)} \right\} dx = 0, \ \forall v \in C_0^\infty (a, b).$$

Integrating by parts and appealing to the fundamental lemma of the calculus of variations (Theorem 1.24) we find

$$\sum_{i=1}^n (-1)^{i+1} \frac{d^i}{dx^i} \left[f_{\xi_i} \left(x, \overline{u}, \cdots, \overline{u}^{(n)} \right) \right] = f_u \left(x, \overline{u}, \cdots, \overline{u}^{(n)} \right). \quad \blacksquare$$

Exercise 2.2.3. (i) Let

$$X_0 = \left\{ v \in C^1 ([a, b]) : v (a) = 0 \right\}.$$

Let $\overline{u} \in X \cap C^2 ([a, b])$ be a minimizer for (P), since $I (\overline{u} + \epsilon v) \ge I (\overline{u}), \ \forall v \in X_0$ and $\forall \epsilon \in \mathbb{R}$, we deduce as in the theorem that

$$\int_a^b \left\{ f_u (x, \overline{u}, \overline{u}') v + f_\xi (x, \overline{u}, \overline{u}') v' \right\} dx = 0, \ \forall v \in X_0.$$

Integrating by parts (bearing in mind that $v (a) = 0$) we find

$$\int_a^b \left\{ \left[f_u - \frac{d}{dx} f_\xi \right] v \right\} dx + f_\xi (b, \overline{u} (b), \overline{u}' (b)) v (b) = 0, \ \forall v \in X_0.$$

Applying the implicit function theorem we can find $\epsilon_0 > 0$ and a function $t \in C^1\left([-\epsilon_0, \epsilon_0]\right)$ with $t(0) = 0$ such that

$$G\left(\epsilon, t\left(\epsilon\right)\right) = 0, \ \forall \epsilon \in [-\epsilon_0, \epsilon_0]$$

which implies that $\overline{u} + \epsilon v + t\left(\epsilon\right) w \in X$. Note also that

$$G_\epsilon\left(\epsilon, t\left(\epsilon\right)\right) + G_h\left(\epsilon, t\left(\epsilon\right)\right) t'\left(\epsilon\right) = 0, \ \forall \epsilon \in [-\epsilon_0, \epsilon_0]$$

and hence

$$t'(0) = -G_\epsilon\left(0, 0\right).$$

Since we know that

$$F(0, 0) \leq F\left(\epsilon, t\left(\epsilon\right)\right), \ \forall \epsilon \in [-\epsilon_0, \epsilon_0]$$

we deduce that

$$F_\epsilon\left(0, 0\right) + F_h\left(0, 0\right) t'\left(0\right) = 0$$

and thus letting $\lambda = F_h\left(0, 0\right)$ be the Lagrange multiplier we find

$$F_\epsilon\left(0, 0\right) - \lambda G_\epsilon\left(0, 0\right) = 0$$

or in other words

$$\int_a^b \left\{[f_\xi\left(x, \overline{u}, \overline{u}'\right) v' + f_u\left(x, \overline{u}, \overline{u}'\right) v] - \lambda\left[g_\xi\left(x, \overline{u}, \overline{u}'\right) v' + g_u\left(x, \overline{u}, \overline{u}'\right) v\right]\right\} dx = 0.$$

Appealing once more to the fundamental lemma of the calculus of variations and to the fact that $v \in C_0^\infty(a, b)$ is arbitrary we get

$$\frac{d}{dx}\left[f_\xi\left(x, \overline{u}, \overline{u}'\right)\right] - f_u\left(x, \overline{u}, \overline{u}'\right) = \lambda\left\{\frac{d}{dx}\left[g_\xi\left(x, \overline{u}, \overline{u}'\right)\right] - g_u\left(x, \overline{u}, \overline{u}'\right)\right\}. \quad \blacksquare$$

Exercise 2.2.5. Let $v \in C_0^1(a, b)$, $\epsilon \in \mathbb{R}$ and set $\varphi\left(\epsilon\right) = I\left(\overline{u} + \epsilon v\right)$. Since $\overline{u}$ is a minimizer of (P) we have $\varphi\left(\epsilon\right) \geq \varphi(0)$, $\forall \epsilon \in \mathbb{R}$, and hence we have that $\varphi'(0) = 0$ (which leads to the Euler-Lagrange equation) and $\varphi''(0) \geq 0$. Computing the last expression we find

$$\int_a^b \left\{f_{uu} v^2 + 2 f_{u\xi} v v' + f_{\xi\xi} v'^2\right\} dx \geq 0, \ \forall v \in C_0^1(a, b).$$

Noting that $2vv' = \left(v^2\right)'$ and recalling that $v(a) = v(b) = 0$, we find

$$\int_a^b \left\{f_{\xi\xi} v'^2 + \left(f_{uu} - \frac{d}{dx} f_{u\xi}\right) v^2\right\} dx \geq 0, \ \forall v \in C_0^1(a, b). \quad \blacksquare$$

We start by observing that for any $\epsilon > 0$ and $u \in X_{\text{piec}}$, we can find $v \in X$ such that

$$\|u - v\|_{W^{1,2}} \leq \epsilon.$$

It is an easy matter (exactly as above) to show that if

$$I(u) = \int_0^1 x \left(u'(x)\right)^2 dx$$

then we can find a constant γ so that

$$0 \leq I(v) \leq I(u) + \gamma\epsilon.$$

Taking the infimum over all elements $v \in X$ and $u \in X_{\text{piec}}$ we get that

$$0 \leq m \leq \gamma\epsilon$$

which is the desired result, since ϵ is arbitrary. ∎

Exercise 2.2.7. Let $u \in C^1([0,1])$ with $u(0) = u(1) = 0$. Invoking Poincaré inequality (cf. Theorem 1.48), we can find a constant $\gamma > 0$ such that

$$\int_0^1 u^2 dx \leq \gamma \int_0^1 u'^2 dx.$$

We hence obtain that $m_\lambda = 0$ if λ is small (more precisely $\lambda^2 \leq 1/\gamma$). Observe that $u_0 \equiv 0$ satisfies $I_\lambda(u_0) = m_\lambda = 0$. Furthermore it is the unique solution of (P_λ) since, by inspection, it is the only solution (if $\lambda^2 < \pi^2$) of the Euler-Lagrange equation

$$\begin{cases} u'' + \lambda^2 u = 0 \\ u(0) = u(1) = 0. \end{cases}$$

The claim then follows. ∎

Exercise 2.2.8. Let

$$X_{\text{piec}} = \left\{ u \in C^1_{\text{piec}}([-1,1]) : u(-1) = 0, \ u(1) = 1 \right\}$$

and

$$\overline{u}(x) = \begin{cases} 0 & \text{if } x \in [-1,0] \\ x & \text{if } x \in (0,1]. \end{cases}$$

It is then obvious to see that

$$\inf_{u \in X_{\text{piec}}} \left\{ I(u) = \int_{-1}^1 f(u(x), u'(x)) \, dx \right\} = I(\overline{u}) = 0.$$

Setting

$$\overline{v}(x) = \frac{A(\beta) - A(\alpha)}{b - a}(x - a) + A(\alpha) \quad \text{and} \quad \overline{u}(x) = A^{-1}(\overline{v}(x))$$

we have from the preceding inequality

$$I(u) \geq (b - a) \left| \frac{A(\beta) - A(\alpha)}{b - a} \right|^p = \int_a^b |\overline{v}'(x)|^p \, dx = I(\overline{u})$$

as claimed. ∎

7.2.2   Second form of the Euler-Lagrange equation

Exercise 2.3.1. Write $f_\xi = (f_{\xi^1}, \cdots, f_{\xi^N})$ and start by the simple observation that for any $u \in C^2([a,b];\mathbb{R}^N)$

$$\frac{d}{dx}[f(x,u,u') - \langle u'; f_\xi(x,u,u') \rangle]$$
$$= f_x(x,u,u') + \sum_{i=1}^N (u^i)' \left[f_{u^i}(x,u,u') - \frac{d}{dx}[f_{\xi^i}(x,u,u')] \right].$$

Since the Euler-Lagrange system (see Exercise 2.2.1) is given by

$$\frac{d}{dx}[f_{\xi^i}(x,\overline{u},\overline{u}')] = f_{u^i}(x,\overline{u},\overline{u}'), \; i = 1, \cdots, N$$

we obtain

$$\frac{d}{dx}[f(x,\overline{u},\overline{u}') - \langle \overline{u}'; f_\xi(x,\overline{u},\overline{u}') \rangle] = f_x(x,\overline{u},\overline{u}'). \quad ∎$$

Exercise 2.3.2. The second form of the Euler-Lagrange equation is

$$0 = \frac{d}{dx}[f(u(x),u'(x)) - u'(x)f_\xi(u(x),u'(x))] = \frac{d}{dx}\left[-u(x) - \frac{1}{2}(u'(x))^2\right]$$
$$= -u'(x) - u''(x)u'(x) = -u'(x)[u''(x) + 1]$$

and it is satisfied by $u \equiv 1$. However $u \equiv 1$ does not verify the Euler-Lagrange equation, which is in the present case

$$u''(x) = -1. \quad ∎$$

In terms of the Hamiltonian, if we let $u_i = (x_i, y_i, z_i)$, $\xi_i = (\xi_i^x, \xi_i^y, \xi_i^z)$ and $v_i = (v_i^x, v_i^y, v_i^z)$, for $i = 1, \cdots, N$, we find

$$H(t, u, v) = \sup_{\xi \in \mathbb{R}^{3N}} \left\{ \sum_{i=1}^{N} \left[\langle v_i; \xi_i \rangle - \frac{1}{2} m_i |\xi_i|^2 \right] + U(t, u) \right\}$$

$$= \sum_{i=1}^{N} \frac{|v_i|^2}{2m_i} + U(t, u).$$

(ii) Note that along the trajectories we have $v_i = m_i u_i'$, i.e.

$$v_i^x = m_i x_i', \quad v_i^y = m_i y_i', \quad v_i^z = m_i z_i'$$

and hence

$$H(t, u, v) = \frac{1}{2} \sum_{i=1}^{N} m_i |u_i'|^2 + U(t, u). \quad \blacksquare$$

Exercise 2.4.3. Although the hypotheses of Theorem 2.11 are not satisfied in the present context, the procedure is exactly the same and leads to the following analysis. The Hamiltonian is

$$H(x, u, v) = \begin{cases} -\sqrt{g(x, u) - v^2} & \text{if } v^2 \leq g(x, u) \\ +\infty & \text{otherwise.} \end{cases}$$

We therefore have, provided $v^2 < g(x, u)$, that

$$\begin{cases} u' = H_v = \dfrac{v}{\sqrt{g(x, u) - v^2}} \\ v' = -H_u = \dfrac{1}{2} \dfrac{g_u}{\sqrt{g(x, u) - v^2}}. \end{cases}$$

We hence obtain that $2vv' = g_u u'$ and thus

$$\left[v^2(x) - g(x, u(x)) \right]' + g_x(x, u(x)) = 0.$$

If $g(x, u) = g(u)$, we get (c being a constant)

$$v^2(x) = c + g(u(x)). \quad \blacksquare$$

7.2.4   Hamilton-Jacobi equation

Exercise 2.5.1. We state without proofs the results (they are similar to the case $N = 1$ and we refer to Gelfand-Fomin [48] page 88, if necessary). Let $H \in$

which implies

$$1 - \frac{\alpha u'(x)}{\sqrt{g(u(x)) - \alpha^2}} = 0.$$

Note that, indeed, any such $u = u(x)$ and

$$v = v(x) = S_u(x, u(x), \alpha) = \sqrt{g(u(x)) - \alpha^2}$$

satisfy

$$\begin{cases} u'(x) = H_v(x, u(x), v(x)) = \frac{\sqrt{g(u(x)) - \alpha^2}}{\alpha} \\ v'(x) = -H_u(x, u(x), v(x)) = \frac{g'(u(x))u'(x)}{2\sqrt{g(u(x)) - \alpha^2}} = \frac{g'(u(x))}{2\alpha} . \end{cases} \blacksquare$$

Exercise 2.5.3. The Hamiltonian is easily seen to be

$$H(u, v) = \frac{v^2}{2a(u)} .$$

The Hamilton-Jacobi equation and its reduced form are given by

$$S_x + \frac{(S_u)^2}{2a(u)} = 0 \quad \text{and} \quad \frac{(S_u^*)^2}{2a(u)} = \frac{\alpha^2}{2} .$$

Therefore, defining A by $A'(u) = \sqrt{a(u)}$, we find

$$S^*(u, \alpha) = \alpha A(u) \quad \text{and} \quad S(x, u, \alpha) = -\frac{\alpha^2}{2}x + \alpha A(u) .$$

Hence, according to Theorem 2.20 (note that $S_{u\alpha} = \sqrt{a(u)} > 0$) the solution is given implicitly by

$$S_\alpha(x, u(x), \alpha) = -\alpha x + A(u(x)) \equiv \beta = \text{constant} .$$

Since A is invertible we find (compare with Exercise 2.2.10)

$$u(x) = A^{-1}(\alpha x + \beta) . \quad \blacksquare$$

7.2.5   Fields theories

Exercise 2.6.1. Let $f \in C^2([a, b] \times \mathbb{R}^N \times \mathbb{R}^N)$, $\alpha, \beta \in \mathbb{R}^N$. Assume that there exists $\Phi \in C^3([a, b] \times \mathbb{R}^N)$ satisfying $\Phi(a, \alpha) = \Phi(b, \beta)$. Suppose also that

$$\widetilde{f}(x, u, \xi) = f(x, u, \xi) + \langle \Phi_u(x, u); \xi \rangle + \Phi_x(x, u)$$

7.3   Chapter 3. Direct methods: existence

7.3.1   The model case: Dirichlet integral

Exercise 3.2.1. The proof is almost completely identical to that of Theorem 3.1; only the first step is slightly different. So let $\{u_\nu\}$ be a minimizing sequence

$$I(u_\nu) \to m = \inf\left\{I(u) : u \in W_0^{1,2}(\Omega)\right\}.$$

Since $I(0) < +\infty$, we have that $m < +\infty$. Consequently we have from Hölder inequality that

$$m + 1 \geq I(u_\nu) = \frac{1}{2}\int_\Omega |\nabla u_\nu|^2\, dx - \int_\Omega h(x)\, u_\nu(x)\, dx$$

$$\geq \int_\Omega \frac{1}{2}|\nabla u_\nu|^2 - \|h\|_{L^2}\|u_\nu\|_{L^2} = \frac{1}{2}\|\nabla u_\nu\|_{L^2}^2 - \|h\|_{L^2}\|u_\nu\|_{L^2}.$$

Using Poincaré inequality (cf. Theorem 1.48) we can find constants (independent of ν) $\gamma_k > 0$, $k = 1, \cdots, 5$, so that

$$m + 1 \geq \gamma_1\|u_\nu\|_{W^{1,2}}^2 - \gamma_2\|u_\nu\|_{W^{1,2}} \geq \gamma_3\|u_\nu\|_{W^{1,2}}^2 - \gamma_4$$

and hence, as wished,

$$\|u_\nu\|_{W^{1,2}} \leq \gamma_5. \quad \blacksquare$$

7.3.2   A general existence theorem

Exercise 3.3.1. As in Exercise 3.2.1 it is the compactness proof in Theorem 3.3 that has to be modified, the remaining part of the proof is essentially unchanged. Let therefore $\{u_\nu\}$ be a minimizing sequence, i.e. $I(u_\nu) \to m$. We have from (H_2) that for ν sufficiently large

$$m + 1 \geq I(u_\nu) \geq \alpha_1\|\nabla u_\nu\|_{L^p}^p - |\alpha_2|\|u_\nu\|_{L^q}^q - |\alpha_3|\,\mathrm{meas}\,\Omega.$$

From now on we denote by $\gamma_k > 0$ constants that are independent of ν. Since by Hölder inequality we have

$$\|u_\nu\|_{L^q}^q = \int_\Omega |u_\nu|^q \leq \left(\int_\Omega |u_\nu|^p\right)^{q/p}\left(\int_\Omega dx\right)^{(p-q)/p} = (\mathrm{meas}\,\Omega)^{(p-q)/p}\|u_\nu\|_{L^p}^q$$

we deduce that we can find constants γ_1 and γ_2 such that

$$m + 1 \geq \alpha_1\|\nabla u_\nu\|_{L^p}^p - \gamma_1\|u_\nu\|_{L^p}^q - \gamma_2$$

$$\geq \alpha_1\|\nabla u_\nu\|_{L^p}^p - \gamma_1\|u_\nu\|_{W^{1,p}}^q - \gamma_2.$$

Since we have from Rellich theorem, that $u_\nu \to \overline{u}$ in L^q, $\forall q \in [1, \infty)$ we obtain the desired convergence.

Case 3: $p < n$. The same argument as in Case 2 leads to the result, the difference being that Rellich theorem now gives $u_\nu \to \overline{u}$ in L^q, $\forall q \in [1, np/(n-p))$. ∎

Exercise 3.3.3. We have here weakened the hypothesis (H_3) in the proof of the theorem. We used this hypothesis only in the lower semicontinuity part of the proof, so let us establish this property under the new condition. Let $u_\nu \rightharpoonup \overline{u}$ in $W^{1,p}((a,b);\mathbb{R}^N)$. Using the convexity of $(u, \xi) \to f(x, u, \xi)$ we find

$$\int_a^b f(x, u_\nu, u'_\nu)\, dx \geq \int_a^b f(x, \overline{u}, \overline{u}')\, dx$$
$$+ \int_a^b [\langle f_u(x, \overline{u}, \overline{u}'); u_\nu - \overline{u}\rangle + \langle f_\xi(x, \overline{u}, \overline{u}'); u'_\nu - \overline{u}'\rangle]\, dx.$$

Since, by Rellich theorem, $u_\nu \to \overline{u}$ in L^∞, to pass to the limit in the second term of the right-hand side of the inequality we need only to have $f_u(x, \overline{u}, \overline{u}') \in L^1$. This is ascertained by the hypothesis

$$|f_u(x, u, \xi)| \leq \beta(1 + |\xi|^p).$$

Similarly to pass to the limit in the last term we need to have $f_\xi(x, \overline{u}, \overline{u}') \in L^{p'}$, $p' = p/(p-1)$; and this is precisely true because of the hypothesis

$$|f_\xi(x, u, \xi)| \leq \beta\left(1 + |\xi|^{p-1}\right). \quad ∎$$

Exercise 3.3.4. We proceed as in Theorem 3.3. Assume that there exist $\overline{u}, \overline{v} \in u_0 + W_0^{1,p}(\Omega)$ so that

$$I(\overline{u}) = I(\overline{v}) = m$$

and we prove that this implies $\overline{u} = \overline{v}$. We obtain, as in the theorem, from the convexity of $(u, \xi) \to f(x, u, \xi)$, that

$$\frac{1}{2}f(x, \overline{u}, \nabla\overline{u}) + \frac{1}{2}f(x, \overline{v}, \nabla\overline{v}) - f\left(x, \frac{\overline{u}+\overline{v}}{2}, \frac{\nabla\overline{u}+\nabla\overline{v}}{2}\right) = 0 \text{ a.e. in } \Omega.$$

We now use the special structure of f to get, since $u \to g(x, u)$ and $\xi \to h(x, \xi)$ are convex, that

$$\frac{1}{2}g(x, \overline{u}) + \frac{1}{2}g(x, \overline{v}) - g\left(x, \frac{\overline{u}+\overline{v}}{2}\right) = 0 \text{ a.e. in } \Omega$$

$$\frac{1}{2}h(x, \nabla\overline{u}) + \frac{1}{2}h(x, \nabla\overline{v}) - h\left(x, \frac{\nabla\overline{u}+\nabla\overline{v}}{2}\right) = 0 \text{ a.e. in } \Omega.$$

Exercise 3.4.2. Use the preceding exercise to deduce the following growth conditions on $g \in C^1\left(\overline{\Omega} \times \mathbb{R}\right)$.

Case 1: $p > n$. No growth condition is imposed on g.

Case 2: $p = n$. There exist $\beta > 0$ and $s_1 \geq 1$ such that

$$\left|g_u\left(x, u\right)\right| \leq \beta\left(1 + |u|^{s_1}\right), \ \forall\left(x, u\right) \in \overline{\Omega} \times \mathbb{R}.$$

Case 3: $p < n$. There exist $\beta > 0$ and $1 \leq s_1 \leq \left(np - n + p\right)/\left(n - p\right)$, so that

$$\left|g_u\left(x, u\right)\right| \leq \beta\left(1 + |u|^{s_1}\right), \ \forall\left(x, u\right) \in \overline{\Omega} \times \mathbb{R}. \quad \blacksquare$$

Exercise 3.4.3. (i) Let ν be an integer and

$$u_\nu\left(x, t\right) = \sin\left(\nu x\right) \sin t.$$

We obviously have $u_\nu \in W_0^{1,2}\left(\Omega\right)$ (in fact $u_\nu \in C^\infty\left(\overline{\Omega}\right)$ and $u_\nu = 0$ on $\partial\Omega$). An elementary computation shows that $\lim_{\nu \to \infty} I\left(u_\nu\right) = -\infty$.

(ii) The second part is elementary.

It is also clear that for the wave equation it is not reasonable to impose an initial condition (at $t = 0$) and a final condition (at $t = \pi$). $\quad \blacksquare$

7.3.4   The vectorial case

Exercise 3.5.1. Let

$$\xi_1 = \begin{pmatrix} 1 & 0 \\ 0 & 0 \end{pmatrix} \quad \text{and} \quad \xi_2 = \begin{pmatrix} 0 & 0 \\ 0 & 1 \end{pmatrix}.$$

We have that both functions are not convex, since

$$\frac{1}{2}f_1\left(\xi_1\right) + \frac{1}{2}f_1\left(\xi_2\right) = \frac{1}{2}\left(\det \xi_1\right)^2 + \frac{1}{2}\left(\det \xi_2\right)^2 = 0 < f_1\left(\frac{1}{2}\xi_1 + \frac{1}{2}\xi_2\right) = \frac{1}{16}$$

$$\frac{1}{2}f_2\left(\xi_1\right) + \frac{1}{2}f_2\left(\xi_2\right) = 1 < f_2\left(\frac{1}{2}\xi_1 + \frac{1}{2}\xi_2\right) = \frac{5}{4}. \quad \blacksquare$$

Exercise 3.5.2. We divide the discussion into two steps.

Step 1. Let $u, v \in C^2\left(\overline{\Omega}; \mathbb{R}^2\right)$ with $u = v$ on $\partial\Omega$. Write

$$u = u\left(x_1, x_2\right) = \left(\varphi\left(x_1, x_2\right), \psi\left(x_1, x_2\right)\right)$$
$$v = v\left(x_1, x_2\right) = \left(\alpha\left(x_1, x_2\right), \beta\left(x_1, x_2\right)\right).$$

Use the fact that

$$\det \nabla u = \varphi_{x_1}\psi_{x_2} - \varphi_{x_2}\psi_{x_1} = \left(\varphi\psi_{x_2}\right)_{x_1} - \left(\varphi\psi_{x_1}\right)_{x_2}$$

Exercise 3.5.3. Let $u \in W^{1,p}\left(\Omega; \mathbb{R}^2\right)$,

$$u\left(x_1, x_2\right) = \left(\varphi\left(x_1, x_2\right), \psi\left(x_1, x_2\right)\right),$$

be a minimizer of (P) and let $v \in C_0^\infty\left(\Omega; \mathbb{R}^2\right)$

$$v\left(x_1, x_2\right) = \left(\alpha\left(x_1, x_2\right), \beta\left(x_1, x_2\right)\right)$$

be arbitrary. Since $I\left(u + \epsilon v\right) \geq I\left(u\right)$ for every ϵ, we must have

$$\left.\frac{d}{d\epsilon} I\left(u + \epsilon v\right)\right|_{\epsilon=0} = 0.$$

Since

$$
\begin{aligned}
& I\left(u + \epsilon v\right) \\
& = \iint_\Omega \left[\left(\varphi_{x_1} + \epsilon\alpha_{x_1}\right)\left(\psi_{x_2} + \epsilon\beta_{x_2}\right) - \left(\varphi_{x_2} + \epsilon\alpha_{x_2}\right)\left(\psi_{x_1} + \epsilon\beta_{x_1}\right)\right] dx_1 dx_2
\end{aligned}
$$

we therefore deduce that

$$\iint_\Omega \left[\left(\psi_{x_2}\alpha_{x_1} - \psi_{x_1}\alpha_{x_2}\right) + \left(\varphi_{x_1}\beta_{x_2} - \varphi_{x_2}\beta_{x_1}\right)\right] dx_1 dx_2 = 0.$$

Integrating by parts, we find that the left-hand side vanishes identically. The result is not surprising in view of Exercise 3.5.2, which shows that $I\left(u\right)$ is in fact constant. ∎

Exercise 3.5.4. The proof is divided into two steps.

Step 1. It is easily proved that the following algebraic inequality holds

$$\left|\det A - \det B\right| \leq \gamma\left(\left|A\right| + \left|B\right|\right)\left|A - B\right|, \ \forall A, B \in \mathbb{R}^{2\times 2}$$

where γ is a constant.

Step 2. We therefore deduce that

$$\left|\det \nabla u - \det \nabla v\right|^{p/2} \leq \gamma^{p/2}\left(\left|\nabla u\right| + \left|\nabla v\right|\right)^{p/2}\left|\nabla u - \nabla v\right|^{p/2}.$$

Hölder inequality implies then

$$
\begin{aligned}
& \iint_\Omega \left|\det \nabla u - \det \nabla v\right|^{p/2} dx_1 dx_2 \\
& \leq \gamma^{p/2}\left(\iint_\Omega \left(\left|\nabla u\right| + \left|\nabla v\right|\right)^p dx_1 dx_2\right)^{1/2}\left(\iint_\Omega \left|\nabla u - \nabla v\right|^p dx_1 dx_2\right)^{1/2}.
\end{aligned}
$$

Passing to the limit and integrating by parts once more we get the claim, namely

$$\lim_{\nu \to \infty} \iint_\Omega \det \nabla u^\nu v \, dx_1 dx_2 = \iint_\Omega \left(\varphi \psi_{x_1} v_{x_2} - \varphi \psi_{x_2} v_{x_1} \right) dx_1 dx_2$$

$$= \iint_\Omega \det \nabla u \, v \, dx_1 dx_2 . \quad \blacksquare$$

Exercise 3.5.6. (i) Let $x = (x_1, x_2)$, we then find

$$\nabla u = \begin{pmatrix} \dfrac{x_2^2}{|x|^3} & -\dfrac{x_1 x_2}{|x|^3} \\[3mm] -\dfrac{x_1 x_2}{|x|^3} & \dfrac{x_1^2}{|x|^3} \end{pmatrix} \Rightarrow |\nabla u|^2 = \dfrac{1}{|x|^2} .$$

We therefore deduce (cf. Exercise 1.4.1) that $u \in L^\infty$ and $u \in W^{1,p}$ provided $p \in [1, 2)$, but, however, $u \notin W^{1,2}$ and $u \notin C^0$.

(ii) Since

$$\iint_\Omega |u^\nu (x) - u (x)|^q \, dx = 2\pi \int_0^1 \frac{r}{(\nu r + 1)^q} dr = \frac{2\pi}{\nu^2} \int_1^{\nu+1} \frac{s - 1}{s^q} ds$$

we deduce that $u^\nu \to u$ in L^q, for every $q \geq 1$; however the convergence $u^\nu \to u$ in L^∞ does not hold. We next show that $u^\nu \rightharpoonup u$ in $W^{1,p}$ if $p \in [1, 2)$. We readily have

$$\nabla u^\nu = \frac{1}{|x| \left(|x| + 1/\nu \right)^2} \begin{pmatrix} x_2^2 + \dfrac{|x|}{\nu} & -x_1 x_2 \\[3mm] -x_1 x_2 & x_1^2 + \dfrac{|x|}{\nu} \end{pmatrix}$$

and thus

$$|\nabla u^\nu| = \frac{\left(|x|^2 + \dfrac{2 |x|}{\nu} + \dfrac{2}{\nu^2} \right)^{1/2}}{\left(|x| + 1/\nu \right)^2} .$$

We therefore find, if $1 \leq p < 2$, that, γ denoting a constant independent of ν,

$$\iint_\Omega |\nabla u^\nu|^p \, dx_1 dx_2 = 2\pi \int_0^1 \frac{\left((r + 1/\nu)^2 + 1/\nu^2 \right)^{p/2}}{(r + 1/\nu)^{2p}} r \, dr$$

$$\leq 2\pi \int_0^1 \frac{2^{p/2} (r + 1/\nu)^p}{(r + 1/\nu)^{2p}} r \, dr = 2^{(2+p)/2} \pi \nu^p \int_0^1 \frac{r \, dr}{(\nu r + 1)^p}$$

$$\leq 2^{(2+p)/2} \pi \nu^{p-2} \int_1^{\nu+1} \frac{(s - 1) \, ds}{s^p} \leq \gamma .$$

ϵ being arbitrary, we have indeed obtained the result. $\blacksquare$

Exercise 3.5.7. **(i)** Start by observing that, in view of Exercise 3.5.2, we have

$$\int_\Omega \det\left(\xi + \nabla\varphi\left(x\right)\right) dx = \det\left(\xi\right) \text{meas}\, \Omega$$

for every $\xi \in \mathbb{R}^{2\times 2}$ and every $\varphi \in W_0^{1,\infty}\left(\Omega; \mathbb{R}^2\right)$. We also trivially have

$$\int_\Omega \left(\xi + \nabla\varphi\left(x\right)\right) dx = \xi \,\text{meas}\, \Omega.$$

We thus deduce, from Jensen inequality that, for every $\xi \in \mathbb{R}^{2\times 2}$ and every $\varphi \in W_0^{1,\infty}\left(\Omega; \mathbb{R}^2\right)$,

$$\int_\Omega f\left(\xi + \nabla\varphi\left(x\right)\right) dx = \int_\Omega F\left(\xi + \nabla\varphi\left(x\right), \det\left(\xi + \nabla\varphi\left(x\right)\right)\right) dx$$
$$\geq F\left(\xi, \det\xi\right) \text{meas}\, \Omega = f\left(\xi\right) \text{meas}\, \Omega$$

as claimed.

(ii) *Step 1.* We start with a preliminary computation. Let $t \in \mathbb{R}$, $\xi \in \mathbb{R}^{2\times 2}$ and $\lambda \in \mathbb{R}^{2\times 2}$ with $\det\lambda = 0$. Observe first that

$$\det\left(\xi + t\lambda\right) = \det\left(\xi\right) + t\left\langle \tilde{\lambda}; \xi \right\rangle$$

where we have denoted

$$\left\langle \tilde{\lambda}; \xi \right\rangle = \lambda_1^1 \xi_2^2 + \lambda_2^2 \xi_1^1 - \lambda_2^1 \xi_1^2 - \lambda_1^2 \xi_2^1.$$

The above observation leads immediately to the following claim. For every $t, s \in \mathbb{R}$, $\xi, \lambda \in \mathbb{R}^{2\times 2}$ with $\det\lambda = 0$ and $\theta \in [0,1]$

$$\xi + \left(\theta t + \left(1 - \theta\right) s\right)\lambda = \theta\left(\xi + t\lambda\right) + \left(1 - \theta\right)\left(\xi + s\lambda\right)$$

$$\det\left[\xi + \left(\theta t + \left(1 - \theta\right) s\right)\lambda\right] = \theta \det\left(\xi + t\lambda\right) + \left(1 - \theta\right)\det\left(\xi + s\lambda\right)$$

or in other words

$$\left(\xi + \left(\theta t + \left(1 - \theta\right) s\right)\lambda, \det\left[\xi + \left(\theta t + \left(1 - \theta\right) s\right)\lambda\right]\right)$$
$$= \theta\left(\xi + t\lambda, \det\left(\xi + t\lambda\right)\right) + \left(1 - \theta\right)\left(\xi + s\lambda, \det\left(\xi + s\lambda\right)\right).$$

Step 2. We have to show that, for every $t, s \in \mathbb{R}$ and $\theta \in [0,1]$,

$$\psi\left(\theta t + \left(1 - \theta\right) s\right) \leq \theta\psi\left(t\right) + \left(1 - \theta\right)\psi\left(s\right).$$

In fact we should need $a = -\infty$ and $b = +\infty$. Recall that in Section 2.2 we already saw that (P) has no minimizer.

(iii) If $f(\xi) = (\xi^2 - 1)^2$, we then find

$$f^{**}(\xi) = \begin{cases} (\xi^2 - 1)^2 & \text{if } |\xi| \geq 1, \\ 0 & \text{if } |\xi| < 1. \end{cases}$$

Therefore if $|\beta - \alpha| \geq 1$ choose in (i) $\lambda = 1/2$ and $a = b = \beta - \alpha$. However if $|\beta - \alpha| < 1$, choose

$$a = 1, \quad b = -1 \quad \text{and} \quad \lambda = (1 + \beta - \alpha)/2.$$

In conclusion, in both cases, we find that problem (P) has $\overline{u}$ as a minimizer. ■

Exercice 3.6.2. Note that if

$$(\overline{P}) \quad \inf \left\{ \overline{I}(u) = \int_\Omega f^{**}(\nabla u(x))\, dx : u \in u_0 + W_0^{1,p}(\Omega) \right\} = \overline{m},$$

then, by Jensen inequality, we have for every $u \in u_0 + W_0^{1,p}(\Omega)$,

$$\int_\Omega f^{**}(\nabla u(x))\, dx \geq \operatorname{meas} \Omega\, f^{**} \left(\frac{1}{\operatorname{meas} \Omega} \int_\Omega \nabla u(x)\, dx \right) = \operatorname{meas} \Omega\, f^{**}(\xi_0).$$

The above computation, coupled with the Relaxation theorem (cf. Theorem 3.28), leads to

$$m = \overline{m} = \overline{I}(u_0) = f^{**}(\xi_0) \operatorname{meas} \Omega. \quad ■$$

Exercise 3.6.3. Let

$$h(\xi) := \alpha_2 |\xi|^p + \alpha_3.$$

Since $\alpha_2 \geq 0$ (in fact $\alpha_2 > 0$), we deduce that h is convex. By hypothesis $h \leq f$ and thus, by definition of f^{**}, we have

$$f(x, u, \xi) \geq f^{**}(x, u, \xi) \geq h(\xi), \quad \forall (x, u, \xi) \in \overline{\Omega} \times \mathbb{R} \times \mathbb{R}^n. \tag{7.17}$$

We can therefore apply Theorem 3.3 to f^{**} to get that $(\overline{P})$ has at least one minimizer $\overline{u} \in u_0 + W_0^{1,p}(\Omega)$.

By Theorem 3.28 (i) we have that there exists a sequence $u_\nu \in u_0 + W_0^{1,p}(\Omega)$ so that

$$u_\nu \to \overline{u} \text{ in } L^p \quad \text{and} \quad I(u_\nu) \to \overline{I}(\overline{u}), \text{ as } \nu \to \infty.$$

From (7.17) and Poincaré inequality we deduce that $\|u_\nu\|_{W^{1,p}}$ is uniformly bounded. Since $p > 1$, we obtain, according to Exercise 1.4.6, that

$$u_\nu \rightharpoonup \overline{u} \text{ in } W^{1,p} \quad \text{and} \quad I(u_\nu) \to \overline{I}(\overline{u}), \text{ as } \nu \to \infty. \quad ■$$

Thus it cannot converge weakly to any u in $W^{1,1}$. Note that the sequence defined on each interval $[k/\nu, (k+1)/\nu]$, $0 \leq k \leq \nu - 1$, by

$$u_\nu(x) = \begin{cases} \sqrt{\nu}\left(x - \frac{k}{\nu}\right) & \text{if } x \in \left[\frac{2k}{2\nu}, \frac{2k+1}{2\nu}\right] \\ \sqrt{\nu}\left(\frac{k+1}{\nu} - x\right) & \text{if } x \in \left(\frac{2k+1}{2\nu}, \frac{2k+2}{2\nu}\right] \end{cases}$$

satisfies $u_\nu \in W_0^{1,1}(0,1)$ with $|u_\nu'| = \sqrt{\nu}$ a.e., $|u_\nu| \leq 1/2\sqrt{\nu}$ and thus

$$u_\nu \to u = 0 \text{ in } L^\infty(0,1), \quad \int_0^1 f(u_\nu'(x))\,dx = e^{-\sqrt{\nu}} \to 0 \text{ as } \nu \to \infty. \quad \blacksquare$$

Exercice 3.6.7. We refer for more details to Marcellini-Sbordone [75], from where the example (ii) below is taken; see also Section 9.2.2 in [31, 2nd edition].

(i) Let $(x_\nu, u_\nu) \to (x, u)$. From Theorem 1.57, we deduce that

$$f^{**}(x_\nu, u_\nu, \xi) \leq \sum_{i=1}^{n+1} \lambda_i f(x_\nu, u_\nu, \xi_i) \quad \text{for every } (\lambda_i, \xi_i) \in \Lambda_\xi$$

where

$$\Lambda_\xi = \left\{ (\lambda_1, \xi_1), \cdots, (\lambda_{n+1}, \xi_{n+1}) \in [0,1] \times \mathbb{R}^n : \sum_{i=1}^{n+1} \lambda_i (1, \xi_i) = (1, \xi) \right\}.$$

Since f is continuous, we get

$$\limsup_{\nu \to \infty} f^{**}(x_\nu, u_\nu, \xi) \leq \sum_{i=1}^{n+1} \lambda_i f(x, u, \xi_i) \quad \text{for every } (\lambda_i, \xi_i) \in \Lambda_\xi.$$

Taking the infimum on all possible choices of $(\lambda_i, \xi_i) \in \Lambda_\xi$, we obtain, appealing once more to Theorem 1.57, that

$$\limsup_{\nu \to \infty} f^{**}(x_\nu, u_\nu, \xi) \leq f^{**}(x, u, \xi)$$

as wished.

(ii) Observe that when $|u| \geq 1$ the function

$$f(u, \xi) = (|\xi| + 1)^{|u|}$$

is convex. If $|u| < 1$, we have, by Theorem 1.57, for any $\nu \geq 1$,

$$1 \leq f^{**}(u, \xi) = f^{**}(u, |\xi|) \leq \frac{|\xi|}{\nu + |\xi|} f(u, \nu + |\xi|) + \frac{\nu}{\nu + |\xi|} f(u, 0).$$

Similarly since $\overline{u} \in W^{1,\infty}$ and $f \in C^\infty$, we can find constant $\gamma_3, \gamma_4 > 0$, such that

$$|f_\xi\left(x, \overline{u}\left(x\right), \overline{u}'\left(x+h\right)\right) - f_\xi\left(x+h, \overline{u}\left(x+h\right), \overline{u}'\left(x+h\right)\right)|$$
$$\leq \gamma_3 \left(|h| + |\overline{u}\left(x+h\right) - \overline{u}\left(x\right)|\right) \leq \gamma_4 |h|.$$

Combining these two inequalities we find

$$|\overline{u}'\left(x+h\right) - \overline{u}'\left(x\right)| \leq \frac{\gamma_2 + \gamma_4}{\gamma_1} |h|$$

as wished; thus $\overline{u} \in W^{2,\infty}\left(a, b\right).$

(ii) Since $\overline{u} \in W^{2,\infty}\left(a, b\right),$ and the Euler-Lagrange equation holds, we get that, for almost every $x \in \left(a, b\right),$

$$\frac{d}{dx}\left[f_\xi\left(x, \overline{u}, \overline{u}'\right)\right] = f_{\xi\xi}\left(x, \overline{u}, \overline{u}'\right)\overline{u}'' + f_{u\xi}\left(x, \overline{u}, \overline{u}'\right)\overline{u}' + f_{x\xi}\left(x, \overline{u}, \overline{u}'\right)$$
$$= f_u\left(x, \overline{u}, \overline{u}'\right).$$

Since (H_1') holds and $\overline{u} \in C^1\left(\left[a, b\right]\right),$ we deduce that there exists $\gamma_5 > 0$ such that

$$f_{\xi\xi}\left(x, \overline{u}\left(x\right), \overline{u}'\left(x\right)\right) \geq \gamma_5 > 0, \ \forall x \in \left[a, b\right].$$

The Euler-Lagrange equation can then be rewritten as

$$\overline{u}'' = \frac{f_u\left(x, \overline{u}, \overline{u}'\right) - f_{x\xi}\left(x, \overline{u}, \overline{u}'\right) - f_{u\xi}\left(x, \overline{u}, \overline{u}'\right)\overline{u}'}{f_{\xi\xi}\left(x, \overline{u}, \overline{u}'\right)}$$

and hence $\overline{u} \in C^2\left(\left[a, b\right]\right).$ Returning to the equation we find that the right-hand side is then C^1, and hence $\overline{u} \in C^3$. Iterating the process we conclude that $\overline{u} \in C^\infty\left(\left[a, b\right]\right),$ as claimed. ∎

Exercise 4.2.2. Note that

$$\overline{u}'\left(x\right) = |x|^{-\frac{2}{7}} x \quad \text{and} \quad \left(\overline{u}'\right)^7 = x^5.$$

We therefore have $\overline{u} \in C^1\left(\left[-1, 1\right]\right),$ more precisely $\overline{u} \in C^{1,5/7}$ and $\left(\overline{u}'\right)^7 \in C^\infty,$ but $\overline{u} \notin C^2$. Moreover $\overline{u}$ satisfies the Euler-Lagrange equation associated to (P_1) and $(P_2),$ namely

$$\left(\left(\overline{u}'\right)^7\right)' = 5x^4.$$

From Theorem 3.3 we deduce that (P_1) and (P_2) have at least one minimizer and from Exercise 3.3.4 that it is unique. From Theorem 3.11 we get that this minimizer is $\overline{u}$. ∎

We have therefore obtained the claim, namely

$$\left| \overline{u}\left(y\right) - \overline{u}\left(x\right) \right| = \left| \int_0^1 \frac{d}{dt} \overline{u}\left(x + t\left(y - x\right)\right) dt \right| \leq \left(2 + \pi\right) \left|y - x\right|.$$

2) It is clear that $\overline{u} \notin C^1\left(\left[-1, 1\right]\right)$ and not even $C_{\text{piec}}^1\left(\left[-1, 1\right]\right)$, in the sense of Definition 1.6, since $\lim_{x \to 0^{\pm}} \overline{u}'\left(x\right)$ does not exist.

3) For $x \neq 0$, we have

$$f\left(x, \overline{u}'\left(x\right)\right) \equiv 0.$$

Since $\overline{u} \in W_0^{1,\infty}\left(-1, 1\right)$ and $f \geq 0$, we therefore get that

$$\inf\left(P\right) = \int_0^1 f\left(x, \overline{u}'\left(x\right)\right) dx = 0.$$

To prove the uniqueness, we just observe that any minimizer should then necessarily satisfy

$$f\left(x, u'\left(x\right)\right) = 0, \quad \text{a.e. in } \left(-1, 1\right)$$

which implies that

$$u'\left(x\right) = 2x \sin\left(\pi/x\right) - \pi \cos\left(\pi/x\right) = \overline{u}'\left(x\right), \quad \text{a.e. in } \left(-1, 1\right).$$

Thus uniqueness is established.   ■

7.4.2   The difference quotient method: interior regularity

Exercise 4.3.1. (i) We require that $G : \mathbb{R}_+ \to \mathbb{R}_+$ is C^2, convex and there exist constants $g_1, g_2, g_3 > 0$ such that, for every $t \geq 0$,

$$0 < g_1 \leq G'\left(t\right) \leq g_2 \quad \text{and} \quad 0 \leq G''\left(t\right) t \leq g_3.$$

Then the function $g\left(\xi\right) = G\left(\left|\xi\right|^2\right)$ satisfies all the hypotheses of Theorem 4.7, since

$$\frac{\partial g}{\partial \xi_i} = 2G'\left(\left|\xi\right|^2\right) \xi_i \quad \text{and} \quad \frac{\partial^2 g}{\partial \xi_i \partial \xi_j} = 2G'\left(\left|\xi\right|^2\right) \delta_{ij} + 4G''\left(\left|\xi\right|^2\right) \xi_i \xi_j.$$

In particular the function $G\left(t\right) = t + \epsilon\left(1 + t\right)^{-1}$ satisfies all the hypotheses, provided $0 \leq \epsilon < 1$, so that we have

$$g\left(\xi\right) = G\left(\left|\xi\right|^2\right) = \left|\xi\right|^2 + \frac{\epsilon}{1 + \left|\xi\right|^2}.$$

7.4.3 The difference quotient method: boundary regularity

Exercise 4.4.1. We change variables and set

$$x = H(y), \quad u(x) = v(H^{-1}(x)), \quad \varphi(x) = \psi(H^{-1}(x))$$

$$y = H^{-1}(x), \quad v(y) = u(H(y)), \quad \psi(y) = \varphi(H(y)).$$

We therefore immediately have

$$u_{x_i}(x) = \sum_{k=1}^{n} v_{y_k}(H^{-1}(x)) \frac{\partial H_k^{-1}}{\partial x_i}(x)$$

$$\varphi_{x_j}(x) = \sum_{l=1}^{n} \psi_{y_l}(H^{-1}(x)) \frac{\partial H_l^{-1}}{\partial x_j}(x)$$

Using the above and the change of variables $x = H(y)$, we obtain

$$\sum_{i,j=1}^{n} \int_U a_{ij}(x) u_{x_i}(x) \varphi_{x_j}(x)\, dx = \sum_{k,l=1}^{n} \int_Q b_{kl}(y) v_{y_k}(y) \psi_{y_l}(y)\, dy$$

where

$$b_{kl}(y) = \sum_{i,j=1}^{n} a_{ij}(H(y)) \frac{\partial H_k^{-1}}{\partial x_i}(H(y)) \frac{\partial H_l^{-1}}{\partial x_j}(H(y)) \det \nabla H(y).$$

We also get, since

$$\sum_{i,j=1}^{n} a_{ij}(x) \lambda_i \lambda_j \geq \alpha |\lambda|^2$$

that (denoting the transpose of a matrix X by X^t)

$$\sum_{k,l=1}^{n} b_{kl}(y) \lambda_k \lambda_l = \det \nabla H \sum_{i,j,k,l=1}^{n} a_{ij} \frac{\partial H_k^{-1}}{\partial x_i} \frac{\partial H_l^{-1}}{\partial x_j} \lambda_k \lambda_l$$

$$= \det \nabla H \sum_{i,j=1}^{n} a_{ij} \left[(\nabla H^{-1})^t \lambda \right]_i \left[(\nabla H^{-1})^t \lambda \right]_j$$

$$\geq \alpha \det \nabla H(y) \left| (\nabla H^{-1}(H(y)))^t \lambda \right|^2.$$

The result

$$\sum_{k,l=1}^{n} b_{kl}(y) \lambda_k \lambda_l \geq \beta |\lambda|^2$$

follows, since H is a regular change of variables. ∎

Exercise 4.5.3. Let $V(r) = |\log r|^\alpha$ and

$$u(x_1, x_2) = x_1 x_2 V(|x|).$$

A direct computation shows that

$$u_{x_1} = x_2 V(|x|) + \frac{x_1^2 x_2}{|x|} V'(|x|) \quad \text{and} \quad u_{x_2} = x_1 V(|x|) + \frac{x_1 x_2^2}{|x|} V'(|x|)$$

while

$$u_{x_1 x_1} = \frac{x_1^3 x_2}{|x|^2} V''(|x|) + \frac{x_1 x_2}{|x|^3} \left(2x_1^2 + 3x_2^2\right) V'(|x|)$$

$$u_{x_2 x_2} = \frac{x_2^3 x_1}{|x|^2} V''(|x|) + \frac{x_1 x_2}{|x|^3} \left(2x_2^2 + 3x_1^2\right) V'(|x|)$$

$$u_{x_1 x_2} = \frac{x_1^2 x_2^2}{|x|^2} V''(|x|) + \frac{x_1^4 + x_1^2 x_2^2 + x_2^4}{|x|^3} V'(|x|) + V(|x|).$$

We therefore get that

$$u_{x_1 x_1}, u_{x_2 x_2} \in C^0\left(\overline{\Omega}\right), \quad u_{x_1 x_2} \notin L^\infty(\Omega). \quad \blacksquare$$

Exercise 4.5.4. Let $V(r) = \log|\log r|$. A direct computation shows that

$$u_{x_1} = \frac{x_1}{|x|} V'(|x|) \quad \text{and} \quad u_{x_2} = \frac{x_2}{|x|} V'(|x|)$$

and therefore

$$u_{x_1 x_1} = \frac{x_1^2}{|x|^2} V''(|x|) + \frac{x_2^2}{|x|^3} V'(|x|)$$

$$u_{x_2 x_2} = \frac{x_2^2}{|x|^2} V''(|x|) + \frac{x_1^2}{|x|^3} V'(|x|)$$

$$u_{x_1 x_2} = \frac{x_1 x_2}{|x|^2} V''(|x|) - \frac{x_1 x_2}{|x|^3} V'(|x|).$$

This leads to

$$\Delta u = V''(|x|) + \frac{V'(|x|)}{|x|} = \frac{-1}{|x|^2 |\log|x||^2} \in L^1(\Omega)$$

while $u_{x_1 x_1}, u_{x_1 x_2}, u_{x_2 x_2} \notin L^1(\Omega)$. Summarizing the results we indeed have that $u \notin W^{2,1}(\Omega)$ while $\Delta u \in L^1(\Omega)$. We also observe (compare with Example 1.33 (ii)) that, trivially, $u \notin L^\infty(\Omega)$ while $u \in W^{1,2}(\Omega)$, since $u \in L^2(\Omega)$ and

$$\iint_\Omega |\nabla u|^2 \, dx = 2\pi \int_0^{1/2} \frac{dr}{r |\log r|^2} = \frac{2\pi}{\log 2}.$$

(iii) Let $\Omega_\epsilon = \left\{ x \in \mathbb{R}^n : \overline{B_\epsilon(x)} \subset \Omega \right\}$. Let $x \in \Omega_\epsilon$, the function $y \rightarrow$ $\varphi_\epsilon(x - y)$ has then its support in Ω since $\operatorname{supp} \varphi_\epsilon \subset \overline{B_\epsilon(0)}$. We therefore have

$$\int_{\mathbb{R}^n} u(y) \varphi_\epsilon(x - y) \, dy = \int_{\mathbb{R}^n} u(x - y) \varphi_\epsilon(y) \, dy$$

$$= \frac{1}{\epsilon^n} \int_{|y| < \epsilon} u(x - y) \psi\left(\frac{|y|}{\epsilon}\right) dy$$

$$= \int_{|z| < 1} u(x - \epsilon z) \psi(|z|) \, dz$$

and thus

$$\int_{\mathbb{R}^n} u(y) \varphi_\epsilon(x - y) \, dy = \int_0^1 \int_{|y| = 1} u(x - \epsilon r y) \psi(r) r^{n-1} dr d\sigma_y.$$

Using (7.20), (7.21) and the above identity, we find

$$\int_{\mathbb{R}^n} u(y) \varphi_\epsilon(x - y) \, dy = u(x).$$

Since $\varphi_\epsilon \in C_0^\infty(\mathbb{R}^n)$, we immediately get that $u \in C^\infty(\Omega_\epsilon)$. Since ϵ is arbitrary, we find that $u \in C^\infty(\Omega)$, as claimed. ∎

7.4.6   Some general results

Exercise 4.7.1. All the hypotheses of Theorem 4.7 are satisfied. Therefore there exists a unique minimizer $u \in u_0 + W_0^{1,2}(\Omega)$ of

$$(P) \quad \inf\left\{ I(u) = \int_\Omega f(\nabla u(x)) \, dx : u \in u_0 + W_0^{1,2}(\Omega) \right\}$$

satisfying $u \in W_{\text{loc}}^{2,2}(\Omega)$ and

$$\sum_{i=1}^n \int_\Omega \frac{\partial f}{\partial \xi_i}(\nabla u) \varphi_{x_i} = 0, \ \forall \varphi \in W_0^{1,2}(\Omega).$$

Let $O \subset \overline{O} \subset \omega \subset \overline{\omega} \subset \Omega$. Fix $k \in \{1, \cdots, n\}$, choose any $\psi \in C_0^\infty(\omega)$ and set in the above equation $\varphi = \psi_{x_k}$. Since $u \in W^{2,2}(\omega)$, we can integrate by parts the equation and get, for every $\psi \in C_0^\infty(\omega)$ (and hence by density in $W_0^{1,2}(\omega)$),

$$0 = \sum_{i=1}^n \int_\omega \frac{\partial}{\partial x_k}\left[\frac{\partial f}{\partial \xi_i}(\nabla u)\right] \psi_{x_i} = \sum_{i,j=1}^n \int_\omega \frac{\partial^2 f}{\partial \xi_i \partial \xi_j}(\nabla u) u_{x_k x_j} \psi_{x_i}.$$

for every $j = 1, \cdots, n$. The Euler-Lagrange equation is then reduced to

$$\int_\Omega \sum_i \left\{ \frac{\partial \overline{u}^j}{\partial x_i} + \left(\sum_k \frac{\partial \overline{u}^k}{\partial x_k} \right) \left(\delta_{ij} + g^{ij} \left(\overline{u} \right) \right) \right\} \varphi^j_{x_i} dx = 0, \ \forall j = 1, \cdots, n \quad (7.23)$$

and for every $\varphi \in W^{1,2}_0 \left(\Omega; \mathbb{R}^n \right)$.

Note that $\overline{u} \in W^{1,2} \left(\Omega; \mathbb{R}^n \right) \cap C^\infty \left(\overline{\Omega} \setminus \{0\}; \mathbb{R}^n \right)$, g^{ij} are bounded and the elements in $\{.\}$ belong to $L^p \left(\Omega \right)$ with $p = 2 > n/\left(n - 1 \right)$ (recall that $n \geq 3$). Therefore, appealing to Exercise 1.4.12, (7.23) will be verified if we can establish (see Step 3) that, for every $x \neq 0$ and for every $j = 1, \cdots, n$,

$$\sum_i \frac{\partial}{\partial x_i} \left\{ \frac{\partial \overline{u}^j}{\partial x_i} + \left(\sum_k \frac{\partial \overline{u}^k}{\partial x_k} \right) \left(\delta_{ij} + g^{ij} \left(\overline{u} \right) \right) \right\} = 0. \quad (7.24)$$

The proof will therefore be complete once (7.22) and (7.24) will be verified.

Step 2. Let us start by observing that

$$\frac{\partial \overline{u}^l}{\partial x_k} = \frac{\partial \overline{u}^k}{\partial x_l} = \frac{\delta_{kl}}{|x|} - \frac{x_k x_l}{|x|^3}.$$

This leads, for every $k = 1, \cdots, n$, to

$$\sum_l \overline{u}^l \frac{\partial \overline{u}^k}{\partial x_l} = \sum_l \overline{u}^l \frac{\partial \overline{u}^l}{\partial x_k} = \sum_l \frac{x_l}{|x|} \left[\frac{\delta_{kl}}{|x|} - \frac{x_k x_l}{|x|^3} \right] = \frac{x_k}{|x|^2} - \frac{x_k}{|x|^4} \sum_l \left(x_l \right)^2 = 0.$$

We therefore have

$$\sum_{k,l} g^{kl} \left(\overline{u} \right) \frac{\partial \overline{u}^l}{\partial x_k} = \frac{4}{\left(n - 2 \right) \left(1 + |\overline{u}|^2 \right)} \sum_k \overline{u}^k \sum_l \overline{u}^l \frac{\partial \overline{u}^l}{\partial x_k} = 0$$

$$\sum_{k,l} g^{kl}_{u^j} \left(\overline{u} \right) \frac{\partial \overline{u}^l}{\partial x_k} = \frac{4}{n - 2} \sum_k \left(\frac{\overline{u}^k}{1 + |\overline{u}|^2} \right)_{u^j} \sum_l \overline{u}^l \frac{\partial \overline{u}^l}{\partial x_k}$$

$$+ \frac{4}{n - 2} \sum_l \left(\overline{u}^l \right)_{u^j} \sum_k \frac{\overline{u}^k}{1 + |\overline{u}|^2} \frac{\partial \overline{u}^l}{\partial x_k}$$

$$= 0$$

establishing (7.22).

Step 3. It remains to prove (7.24). Note first that

$$\sum_k \frac{\partial \overline{u}^k}{\partial x_k} = \sum_k \left[\frac{1}{|x|} - \frac{\left(x_k \right)^2}{|x|^3} \right] = \frac{n - 1}{|x|}.$$

and hence

$$E = a^2 + y^2, \ F = 0, \ G = 1, \ L = N = 0, \ M = \frac{a}{\sqrt{a^2 + y^2}}$$

which leads to $H = 0$, as wished.

(ii) A straightforward computation gives

$$v_x = \left(1 - x^2 + y^2, -2xy, 2x\right), \ v_y = \left(2xy, -1 + y^2 - x^2, -2y\right)$$

$$e_3 = \frac{\left(2x, 2y, x^2 + y^2 - 1\right)}{\left(1 + x^2 + y^2\right)}$$

$$v_{xx} = (-2x, \ -2y, 2), \ v_{xy} = (2y, -2x, 0), \ v_{yy} = (2x, 2y, -2)$$

and hence

$$E = G = \left(1 + x^2 + y^2\right)^2, \ F = 0, \ L = -2, \ N = 2, \ M = 0$$

which shows that, indeed, $H = 0$. ∎

Exercise 5.2.3. (i) Since $|v_x \times v_y|^2 = w^2 \left(1 + w'^2\right)$, we obtain the result.

(ii) Observe that the function

$$f\left(w, \xi\right) = w\sqrt{1 + \xi^2}$$

is not convex over $(0, +\infty) \times \mathbb{R}$; although the function $\xi \to f\left(w, \xi\right)$ is strictly convex, whenever $w > 0$. We therefore only give necessary conditions for existence of minimizers of (P_α) and hence we write the Euler-Lagrange equation associated to (P_α), namely

$$\frac{d}{dx}\left[f_\xi\left(w, w'\right)\right] = f_w\left(w, w'\right) \quad \Leftrightarrow \quad \frac{d}{dx}\left[\frac{ww'}{\sqrt{1 + w'^2}}\right] = \sqrt{1 + w'^2}. \qquad (7.25)$$

Invoking Theorem 2.8, we find that any minimizer w of (P_α) satisfies

$$\frac{d}{dx}\left[f\left(w, w'\right) - w' f_\xi\left(w, w'\right)\right] = 0 \quad \Leftrightarrow \quad \frac{d}{dx}\left[\frac{w}{\sqrt{1 + w'^2}}\right] = 0$$

which implies, if we let $a > 0$ be a constant,

$$w'^2 = \frac{w^2}{a^2} - 1. \qquad (7.26)$$

Before proceeding further, let us observe the following facts.

1) The function $w \equiv a$ is a solution of (7.26) but not of (7.25) and therefore it is irrelevant for our analysis.

We therefore get

$$E^\epsilon = |v_x + \epsilon\varphi e_{3x} + \epsilon\varphi_x e_3|^2 = E + 2\epsilon\left[\varphi_x \langle v_x; e_3 \rangle + \varphi \langle v_x; e_{3x} \rangle\right] + O\left(\epsilon^2\right)$$

$$F^\epsilon = F + \epsilon\left[\varphi_x \langle v_y; e_3 \rangle + \varphi_y \langle v_x; e_3 \rangle + \varphi \langle v_y; e_{3x} \rangle + \varphi \langle v_x; e_{3y} \rangle\right] + O\left(\epsilon^2\right)$$

$$G^\epsilon = G + 2\epsilon\left[\varphi_y \langle v_y; e_3 \rangle + \varphi \langle v_y; e_{3y} \rangle\right] + O\left(\epsilon^2\right)$$

where $O\left(t\right)$ stands for a function f so that $|f\left(t\right)/t|$ is bounded in a neighborhood of $t = 0$. Appealing to the definition of L, M, N, Exercise 5.2.1 and to the fact that $\langle v_x; e_3 \rangle = \langle v_y; e_3 \rangle = 0$, we obtain

$$\begin{aligned}
E^\epsilon G^\epsilon - \left(F^\epsilon\right)^2 &= \left(E - 2\epsilon L\varphi\right)\left(G - 2\epsilon\varphi N\right) - \left(F - 2\epsilon\varphi M\right)^2 + O\left(\epsilon^2\right) \\
&= EG - F^2 - 2\epsilon\varphi\left[EN - 2FM + GL\right] + O\left(\epsilon^2\right) \\
&= \left(EG - F^2\right)\left[1 - 4\epsilon\varphi H\right] + O\left(\epsilon^2\right).
\end{aligned}$$

We therefore conclude that

$$\left|v_x^\epsilon \times v_y^\epsilon\right| = |v_x \times v_y|\left(1 - 2\epsilon\varphi H\right) + O\left(\epsilon^2\right)$$

and hence

$$\text{Area}\left(\Sigma^\epsilon\right) = \text{Area}\left(\Sigma_0\right) - 2\epsilon \iint_\Omega \varphi H |v_x \times v_y|\, dx dy + O\left(\epsilon^2\right). \tag{7.28}$$

Using (7.27) and (7.28) (i.e., we perform the derivative with respect to ϵ) we get

$$\iint_\Omega \varphi H |v_x \times v_y|\, dx dy = 0, \ \forall\varphi \in C_0^\infty\left(\Omega\right).$$

Since $|v_x \times v_y| > 0$ (due to the fact that Σ_0 is a regular surface), we deduce from the fundamental lemma of the calculus of variations (Theorem 1.24) that $H = 0$. ∎

7.5.2 The Douglas-Courant-Tonelli method

Exercise 5.3.1. We have

$$w_x = v_\lambda \lambda_x + v_\mu \mu_x, \ w_y = v_\lambda \lambda_y + v_\mu \mu_y$$

and thus

$$\begin{aligned}
|w_x|^2 &= |v_\lambda|^2 \lambda_x^2 + 2\lambda_x \mu_x \langle v_\lambda; v_\mu \rangle + \mu_x^2 |v_\mu|^2 \\
|w_y|^2 &= |v_\lambda|^2 \lambda_y^2 + 2\lambda_y \mu_y \langle v_\lambda; v_\mu \rangle + \mu_y^2 |v_\mu|^2.
\end{aligned}$$

7.6   Chapter 6. Isoperimetric inequality

7.6.1   The case of dimension 2

Exercise 6.2.1. One can consult Hardy-Littlewood-Polya [58], page 185, for more details. Let $u \in X$ where

$$X = \left\{ u \in W^{1,2}(-1,1) : u(-1) = u(1) \text{ with } \int_{-1}^{1} u = 0 \right\}.$$

Define

$$z(x) = u(x+1) - u(x)$$

and note that $z(-1) = -z(0)$, since $u(-1) = u(1)$. We deduce that we can find $\alpha \in (-1,0]$ so that $z(\alpha) = 0$, which means that $u(\alpha+1) = u(\alpha)$. We denote this common value by a (i.e. $u(\alpha+1) = u(\alpha) = a$). Since $u \in W^{1,2}(-1,1)$ it is easy to see that the function

$$v(x) = (u(x) - a)^2 \cot[\pi(x - \alpha)]$$

vanishes at $x = \alpha$ and $x = \alpha+1$ (this follows from Hölder inequality, see Exercise 1.4.3). We therefore have (recalling that $u(-1) = u(1)$ and $\cot[\pi(1 - \alpha)] = \cot[\pi(-1 - \alpha)]$)

$$\int_{-1}^{1} \left\{ u'^2 - \pi^2 (u - a)^2 - (u' - \pi(u - a) \cot[\pi(x - \alpha)])^2 \right\} dx$$

$$= \pi \left[(u(x) - a)^2 \cot[\pi(x - \alpha)] \right]_{-1}^{1} = 0.$$

Since $\int_{-1}^{1} u = 0$, we get from the above identity that

$$\int_{-1}^{1} (u'^2 - \pi^2 u^2) \, dx = 2\pi^2 a^2 + \int_{-1}^{1} (u' - \pi(u - a) \cot[\pi(x - \alpha)])^2 \quad dx$$

and hence Wirtinger inequality follows. Moreover we have equality in Wirtinger inequality if and only if $a = 0$ and, c denoting a constant,

$$u' = \pi u \cot[\pi(x - \alpha)] \quad \Leftrightarrow \quad u = c \sin[\pi(x - \alpha)]. \quad \blacksquare$$

Exercise 6.2.2. Since the minimum in (P) is attained by $u \in X$, we have, for any $v \in X \cap C_0^\infty(-1,1)$ and any $\epsilon \in \mathbb{R}$, that

$$I(u + \epsilon v) \geq I(u).$$

So that in these new notations

$$L(u,v) = L(a,b) \quad \text{and} \quad M(u,v) = M(a,b).$$

We next let

$$O = \left\{ x \in (a,b) : u'^2(x) + v'^2(x) > 0 \right\}.$$

The case where $O = (a,b)$ has been considered in Step 1 of Theorem 6.4. If O is empty the result is trivial, so we assume from now on that this is not the case. Since the functions u' and v' are continuous, the set O is open. We can then find (see Theorem 6.59 in [60] or Theorem 9 of Chapter 1 in [37])

$$a \leq a_i < b_i < a_{i+1} < b_{i+1} \leq b, \quad \forall i \geq 1$$

$$O = \overset{\infty}{\underset{i=1}{\cup}} (a_i, b_i).$$

In the complement of O, O^c, we have $u'^2 + v'^2 = 0$, and hence

$$L(b_i, a_{i+1}) = M(b_i, a_{i+1}) = 0. \tag{7.30}$$

Step 2. We then change the parametrization on every (a_i, b_i). We choose a multiple of the arc length, namely

$$\begin{cases} y = \eta(x) = -1 + 2\dfrac{L(a,x)}{L(a,b)} \\ \varphi(y) = u\left(\eta^{-1}(y)\right) \quad \text{and} \quad \psi(y) = v\left(\eta^{-1}(y)\right). \end{cases}$$

Note that this is well defined, since $(a_i, b_i) \subset O$. We then let

$$\alpha_i = -1 + 2\frac{L(a,a_i)}{L(a,b)} \quad \text{and} \quad \beta_i = -1 + 2\frac{L(a,b_i)}{L(a,b)}$$

so that

$$\beta_i - \alpha_i = 2\frac{L(a_i,b_i)}{L(a,b)}.$$

Furthermore, since $L(b_i, a_{i+1}) = 0$, we get

$$\beta_i = \alpha_{i+1} \quad \text{and} \quad \overset{\infty}{\underset{i=1}{\cup}} [\alpha_i, \beta_i] = [-1,1].$$

We also easily find that, for $y \in (\alpha_i, \beta_i)$,

$$\sqrt{\varphi'^2(y) + \psi'^2(y)} = \frac{L(a,b)}{2} = \frac{L(a_i,b_i)}{\beta_i - \alpha_i}$$

$$\varphi(\alpha_i) = u(a_i), \ \psi(\alpha_i) = v(a_i), \ \varphi(\beta_i) = u(b_i), \ \psi(\beta_i) = v(b_i).$$

Exercise 6.3.2. **(i)** We adopt the same notations as those of Exercise 5.2.4. By hypothesis there exist a bounded smooth domain $\Omega \subset \mathbb{R}^2$ and a map $v \in C^2(\overline{\Omega}; \mathbb{R}^3)$ $(v = v(x, y),$ with $v_x \times v_y \neq 0$ in $\overline{\Omega})$ so that $\partial A_0 = v(\overline{\Omega})$.

From the divergence theorem it follows that

$$M(A_0) = \frac{1}{3} \iint_\Omega \langle v; v_x \times v_y \rangle \, dx dy. \tag{7.33}$$

Let then $\epsilon \in \mathbb{R}$, $\varphi \in C_0^\infty(\Omega)$ and

$$v^\epsilon(x, y) = v(x, y) + \epsilon \varphi(x, y) e_3$$

where $e_3 = (v_x \times v_y) / |v_x \times v_y|$.

We next consider

$$\partial A^\epsilon = \left\{ v^\epsilon(x, y) : (x, y) \in \overline{\Omega} \right\} = v^\epsilon(\overline{\Omega}).$$

We have to evaluate $M(A^\epsilon)$ and we start by computing

$$
\begin{aligned}
v_x^\epsilon \times v_y^\epsilon &= (v_x + \epsilon(\varphi_x e_3 + \varphi e_{3x})) \times (v_y + \epsilon(\varphi_y e_3 + \varphi e_{3y})) \\
&= v_x \times v_y + \epsilon[\varphi(e_{3x} \times v_y + v_x \times e_{3y})] \\
&\quad + \epsilon[\varphi_x e_3 \times v_y + \varphi_y v_x \times e_3] + O(\epsilon^2)
\end{aligned}
$$

(where $O(t)$ stands for a function f so that $|f(t)/t|$ is bounded in a neighborhood of $t = 0$) which leads to

$$
\begin{aligned}
\langle v^\epsilon; v_x^\epsilon \times v_y^\epsilon \rangle &= \langle v + \epsilon \varphi e_3; v_x^\epsilon \times v_y^\epsilon \rangle \\
&= \langle v; v_x \times v_y \rangle + \epsilon \varphi \langle e_3; v_x \times v_y \rangle + \epsilon \langle v; \varphi(e_{3x} \times v_y + v_x \times e_{3y}) \rangle \\
&\quad + \epsilon \langle v; \varphi_x e_3 \times v_y + \varphi_y v_x \times e_3 \rangle + O(\epsilon^2).
\end{aligned}
$$

Observing that

$$\langle e_3; v_x \times v_y \rangle = |v_x \times v_y|$$

and returning to (7.33), we get after integration by parts that (recalling that $\varphi = 0$ on $\partial \Omega$)

$$
\begin{aligned}
M(A^\epsilon) - M(A_0) &= \frac{\epsilon}{3} \iint_\Omega \varphi \{ |v_x \times v_y| + \langle v; e_{3x} \times v_y + v_x \times e_{3y} \rangle \\
&\quad - (\langle v; e_3 \times v_y \rangle)_x - (\langle v; v_x \times e_3 \rangle)_y \} \, dx dy + O(\epsilon^2) \\
&= \frac{\epsilon}{3} \iint_\Omega \varphi \{ |v_x \times v_y| - \langle v_x; e_3 \times v_y \rangle \\
&\quad - \langle v_y; v_x \times e_3 \rangle \} \, dx dy + O(\epsilon^2).
\end{aligned}
$$

[14] Brézis H., *Analyse fonctionnelle, théorie et applications*, Masson, Paris, 1983.

[15] Buttazzo G., *Semicontinuity, relaxation and integral represention in the calculus of variations*, Pitman, Longman, London, 1989.

[16] Buttazzo G., Ferone V. and Kawohl B., Minimum problems over sets of concave functions and related questions, *Math. Nachr.* **173** (1995), 71-89.

[17] Buttazzo G., Giaquinta M. and Hildebrandt S., *One dimensional variational problems*, Oxford University Press, Oxford, 1998.

[18] Buttazzo G. and Kawohl B., On Newton's problem of minimal resistance, *Math. Intell.* **15** (1992), 7-12.

[19] Carathéodory C., *Calculus of variations and partial differential equations of the first order*, Holden Day, San Francisco, 1965.

[20] Cesari L., *Optimization - Theory and applications*, Springer, New York, 1983.

[21] Chern S.S., An elementary proof of the existence of isothermal parameters on a surface, *Proc. Amer. Math. Soc.*, **6** (1955), 771-782.

[22] Ciarlet P., *Mathematical elasticity, Volume 1, Three dimensional elasticity*, North Holland, Amsterdam, 1988.

[23] Clarke F.H., *Optimization and nonsmooth analysis*, Wiley, New York, 1983.

[24] Courant R., *Dirichlet's principle, conformal mapping and minimal surfaces*, Interscience, New York, 1950.

[25] Courant R., *Calculus of variations*, Courant Institute Publications, New York, 1962.

[26] Courant R. and Hilbert D., *Methods of mathematical physics*, Wiley, New York, 1966.

[27] Crandall M.G., Ishii H. and Lions P.L., User's guide to viscosity solutions of second order partial differential equations, *Bull. Amer. Math. Soc.* **27** (1992), 1-67.

[28] Croce G. and Dacorogna B., On a generalized Wirtinger inequality, *Discrete Contin. Dyn. Syst. Ser. A*, **9** (2003), 1329-1341.

[44] Evans L.C., *Partial differential equations*, Amer. Math. Soc., Providence, 1998.

[45] Evans L.C. and Gariepy R.F., *Measure theory and fine properties of functions*, Studies in Advanced Math., CRC Press, Boca Raton, 1992.

[46] Federer H., *Geometric measure theory*, Springer, Berlin, 1969.

[47] Folland G.B., *Introduction to partial differential equations*, Princeton University Press, Princeton, 1976.

[48] Gelfand I.M. and Fomin S.V., *Calculus of variations*, Prentice-Hall, Englewood, 1963.

[49] Giaquinta M., *Multiple integrals in the calculus of variations and nonlinear elliptic systems*, Princeton University Press, Princeton, 1983.

[50] Giaquinta M. and Hildebrandt S., *Calculus of variations I and II*, Springer, Berlin, 1996.

[51] Gilbarg D. and Trudinger N.S., *Elliptic partial differential equations of second order*, Springer, Berlin, 1977.

[52] Giusti E., *Minimal surfaces and functions of bounded variations*, Birkhäuser, Boston, 1984.

[53] Giusti E., *Metodi diretti del calcolo delle variazioni*, Unione Matematica Italiana, Bologna, 1994. English translation: *Direct methods in the calculus of variations*, World Scientific, Singapore, 2003.

[54] Giusti E. and Miranda M., Un esempio di soluzioni discontinue per un problema di minimo relativo ad un integrale regolare del calcolo delle variazioni, *Boll. Un. Mat. Ital.* **1** (1968), 219-226.

[55] Goldstine H.H., *A history of the calculus of variations from the 17th to the 19th century*, Springer, Berlin, 1980.

[56] Hadamard J., Sur quelques questions du calcul des variations, *Bulletin Société Math. de France* **33** (1905), 73-80.

[57] Hadamard J., *Leçons sur le calcul des variations*, Hermann, Paris, 1910.

[58] Hardy G.H., Littlewood J.E. and Polya G., *Inequalities*, Cambridge University Press, Cambridge, 1961.

[59] Hestenes M.R., *Calculus of variations and optimal control theory*, Wiley, New York, 1966.

[75] Marcellini P. and Sbordone C., Semicontinuity problems in the calculus of variations, *Nonlinear Anal.*, **4** (1980), 241-257.

[76] Mawhin J. and Willem M., *Critical point theory and Hamiltonian systems*, Springer, Berlin, 1989.

[77] Monna A.F., *Dirichlet's principle: a mathematical comedy of errors and its influence on the development of analysis*, Oosthoeck, Utrecht, 1975.

[78] Morrey C.B., Quasiconvexity and the lower semicontinuity of multiple integrals, *Pacific J. Math.* **2** (1952), 25-53.

[79] Morrey C.B., *Multiple integrals in the calculus of variations*, Springer, Berlin, 1966.

[80] Morse M., *The calculus of variations in the large*, Amer. Math Soc., New York, 1934.

[81] Necas J., *Les méthodes directes en théorie des équations elliptiques*, Masson, Paris, 1967.

[82] Nitsche J.C., *Lecture on minimal surfaces*, Cambridge University Press, Cambridge, 1989.

[83] Ornstein D., A non-inequality for differential operators in the L^1 norm, *Arch. Rational Mech. Anal.* **11** (1962), 40-49.

[84] Osserman R., *A survey on minimal surfaces*, Van Nostrand, New York, 1969.

[85] Osserman R., The isoperimetric inequality, *Bull. Amer. Math. Soc.* **84** (1978), 1182-1238.

[86] Pars L., *An introduction to the calculus of variations*, Heinemann, London, 1962.

[87] Payne L., Isoperimetric inequalities and their applications, *SIAM Rev.* **9** (1967), 453-488.

[88] Pisier G., *The volume of convex bodies and Banach space geometry*, Cambridge University Press, Cambridge, 1989.

[89] Polya G. and Szegö G., *Isoperimetric inequalities in mathematical physics*, Princeton University Press, Princeton, 1951.

[90] Porter T.I., A history of the classical isoperimetric problem, in *Contributions to the calculus of variations (1931-1932)*, ed. by Bliss G.A. and Graves L.M., University of Chicago Press, Chicago, 1933.